Creating and Using Instructionally Supportive Assessments in *NGSS* Classrooms

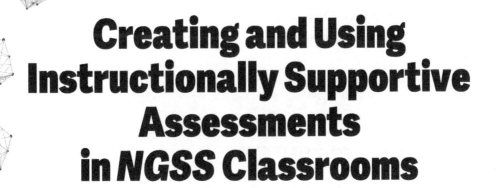

Creating and Using Instructionally Supportive Assessments in *NGSS* Classrooms

EDITORS

Christopher J. Harris
Joseph Krajcik
James W. Pellegrino

FOREWORD BY
Stephen L. Pruitt

National Science Teaching Association

RICHMOND, VIRGINIA

Cathy Iammartino, Director of Publications and Digital Initiatives

PRINTING AND PRODUCTION
Colton Gigot, Senior Production Manager

DESIGN, PRODUCTION, AND PROJECT MANAGEMENT
KTD+ Education Group

National Science Teaching Association
Erika C. Shugart, PhD, Executive Director

405 E. Laburnum Ave., Ste. 3, Richmond, VA 23222
NSTA.org/store
For customer service inquiries, please call 800-277-5300.

Cover images sourced from Shutterstock: Halfpoint (left), PeopleImages.com - Yuri A (middle), Robert Kneschke (right), CoreDESIGN (gear illustrations).

ISBN 978-1-68140-704-3

A catalog record of this book is available from the Library of Congress.

Table of Contents

Contributors

Nonye Alozie
SRI International
Menlo Park, California

Chanyah Dahsah
Srinakharinwirot University
Bangkok, Thailand

Daniel Damelin
Concord Consortium
Concord, Massachusetts

Brian D. Gane
University of Kansas
Lawrence, Kansas

Diksha Gaur
University of Illinois Chicago
Chicago, Illinois

Christopher J. Harris
WestEd
San Francisco, California

Joseph Krajcik
Michigan State University
East Lansing, Michigan

Jane Lee
Michigan State University
East Lansing, Michigan

Krystal Madden
University of Illinois Chicago
Chicago, Illinois

Kevin W. McElhaney
Digital Promise
Redwood City, California

Consuelo J. Morales
Michigan State University
East Lansing, Michigan

James W. Pellegrino
University of Illinois Chicago
Chicago, Illinois

Phyllis Pennock
Great Minds PBC
Farmington Hills, Michigan

Stephen L. Pruitt
Southern Regional Education Board
Atlanta, Georgia

Samuel Severance
Northern Arizona University
Flagstaff, Arizona

Sania Zahra Zaidi
University of Illinois Chicago
Chicago, Illinois

Foreword

ASSESSMENT. Unfortunately, this word has reached new heights of disdain among much of the education community—and the broader public. But let's take a step back from that. Teachers have always given assessments. They are commonplace in other high-performing countries. They have the potential to promote student learning. The term, however, became connected solely to federal and state accountability structures, and state tests often reduced the curriculum to rote memorization. The reality is that assessment, particularly classroom-based assessment, has always been important to instruction. What needs to change is for good assessment to become part of good instruction. This means that both assessment and instruction fully reflect how a scientifically literate person thinks and makes decisions.

From the beginning, the development of the *Next Generation Science Standards* (*NGSS*) were intended to make learning science an active experience, as well as to change how states and districts approach high-stakes assessments. The idea that asking fact-based questions did not portray science in an appropriate manner was a key discussion point during the development of *A Framework for K–12 Science Education* from the National Research Council and the *NGSS*. In one of our first meetings, we clearly stated the goal "to perturb the assessment system." Out of these conversations, we discussed terms and concepts like "performance expectations," "phenomena-based instruction," and "assessment boundaries." We envisioned the *NGSS* to represent, to the extent possible in standards, what science should be: the accumulation and application of concepts across disciplines to address phenomena that students experience or observe. Frankly, this did more than "perturb the system." It caused a complete overhaul. The assessment of scientific and engineering practices, crosscutting concepts, and disciplinary core ideas as one to explain phenomena or develop solutions was new and progressive, and very difficult.

Since the release of *NGSS*, the authors have worked diligently to develop classroom assessments that fully embody science and the *NGSS*. *Creating and Using Instructionally Supportive Assessments in NGSS Classrooms* brings together their work and presents it to educators as a handbook for how we should approach science assessment. The *Next Generation Science Assessment (NGSA)* design process provides a step-by-step approach to help educators change classroom assessment to one that will not only meet the standards but also help students develop higher and more critical thinking skills. Over the next decade, the workforce is going to be based primarily on the ability to use Science, Technology, Engineering, and Mathematics (STEM) ideas. The days of science serving only students who want to work in lab coats are coming to a close. STEM careers are being redefined to include construction, service, welding, transportation, and more. We can no longer believe that having a student pick the right picture representing "anaphase" is enough to learn science. We need assessments that give teachers the information they need to shift instruction, thus allowing students to become STEM thinkers and "doers." The authors have done the hard work of making three-dimensional assessment easier for educators to make this shift. I applaud and appreciate their efforts. I hope all who read this book will give this approach a real try. It will challenge teachers in using and developing assessments aligned with *NGSS*, but nothing worth doing comes easy. There are no magic potions for education. In fact, magic potions work only for wizards. Now let's get to work on behalf of our students.

Stephen L. Pruitt, Ph.D.
President, Southern Regional Education Board
Coordinator of the Development of the Next Generation Science Standards

Preface

I N THE LAST TWENTY YEARS, we have learned a great deal about the teaching and learning of science. This knowledge includes the high value of orchestrating instructional experiences that provide opportunities for students to make sense of compelling phenomena and design solutions to challenging ill-structured problems. When students have these opportunities, they become immersed in the practices and ways of reasoning in science and engineering. In this process, they deepen their disciplinary knowledge and their proficiency in using their knowledge in varying situations. They also learn that science is an active quest to figure out the *how* and *why* of the world around us and that engineering is a means to design solutions to problems that relate to our everyday lives. More than a decade ago, *A Framework for K–12 Science Education* (*Framework*) took into consideration this empowering principle of learning to provide a new vision of science education in which learners need to make sense of phenomena and design solutions using disciplinary core ideas, crosscutting concepts, and science and engineering practices. This vision shifts the focus on science instruction from students learning a static body of knowledge to students using science ideas and practices to generate explanations of phenomena and solve challenging problems based on evidence and scientific reasoning.

With the release of the *Framework* and soon thereafter the *Next Generation Science Standards* (*NGSS*), science education entered an exciting new era where notions of knowledge-in-use, phenomenon-based instruction, and three-dimensional learning began to take hold. New science standards that mirrored the three-dimensional learning goals of the *NGSS*, referred to as *performance expectations*, also gained traction in many corners of the United States. At the same time, a small interdisciplinary group of experts met to explore the implications of this remarkable new vision for science assessment. We realized that science teachers would need innovative assessment tasks and accompanying resources that they could use in their classrooms to support their teaching and enhance their students' three-dimensional learning. To address this pressing need, we formed a cross-institutional group that soon became known as the *Next Generation Science Assessment (NGSA)* collaborative. The *NGSA* collaborative brought together an interdisciplinary team with founding members from the Concord Consortium, CREATE for STEM Institute at Michigan State University, the Learning Sciences Research Institute at University of Illinois Chicago, and the Center for Technology in Learning at SRI International. The collaborative later reconfigured itself to include the Science and Engineering Education Research team at WestEd. The *NGSA* collaborative grew to include a wide range of people with collective expertise in:

- science disciplinary knowledge and practice;
- science teaching and learning;
- classroom-based assessment and educational measurement;
- technology-enhanced instruction and assessment;
- diversity, equity, and inclusion in science education; and
- science teacher professional learning.

As we organized our efforts, we also came to the realization that there was no clear systematic guidance on how to design assessment tasks for classroom use that would help teachers to assess their students in building toward the performance expectations of the *NGSS*. We knew that a principled approach would be paramount if we wanted to equip ourselves and other assessment designers—whether these designers are classroom teachers, state and district educational leaders, or curriculum and assessment developers— to create classroom-based tasks that truly capture three-dimensional performance. Early on, we turned to evidence-centered design (ECD) as a foundation for building our *NGSA* design process. ECD emphasizes the value of starting with a learning goal and determining the evidence that you would look for to make a judgment about students' performance of that learning goal, then specifying the features of tasks that will best bring out the evidence of performance. It is a straightforward, yet powerful idea: assessment tasks should measure what matters and what matters should be drawn from well-specified learning goals. ECD has been used in a wide range of assessment design contexts; however, when we began our work, no one yet had tried to use ECD to design three-dimensional assessment tasks for science classroom settings.

With ECD providing our framework, we set out to develop a comprehensive solution that would address four pressing problems in classroom-based science assessment:

1. *How can we design assessment tasks* in which students make sense of phenomena or design solutions to problems using the three dimensions of scientific knowledge?

2. *How can we use performance expectations to construct assessment tasks* so that teachers can gauge students' progress toward achieving them?

3. *How can we establish alignment of tasks* with performance expectations within the design process?

4. *How can we ensure that tasks are equitable and inclusive for all students* so that they can demonstrate three-dimensional learning?

Our solution is the *NGSA* design process that we describe in this book and that can be used to develop instructionally supportive tasks that assess for three-dimensional learning. The collaborative team developed and refined the design process over multiple interactive cycles where, each time, we would put the design process to the test by using it to create tasks for classrooms, carefully evaluate the process and product, and then revise the process based on how well it performed. During early cycles, we documented each step and grappled with how to clearly describe how we begin with performance expectations and end with three-dimensional tasks and rubrics. In later cycles, we continually asked, *How can the process be improved?* Along the way, we shared our iterations with science educators and science assessment experts to invite feedback on the feasibility of the process, identify gaps and issues, and refine the design model so that it could be easily grasped and taken up. Over the years, we have refined the assessment design process so that it can be utilized by classroom teachers, district science leaders, science assessment developers, and others who want a step-by-step process to create classroom-ready assessment tasks.

Today, the *NGSA* design process has been used to develop assessment tasks across the wide range of science domains and for use with students from the earliest elementary grades upward to undergraduate science courses. Our process is not just another set of guidelines; if taken up and thoughtfully used, each phase of the process provides an opportunity for learning that will increase your knowledge for three-dimensional instruction and assessment. Becoming familiar with the whole process will enable you to develop the knowledge and capabilities to design a variety of tasks that can provide a picture of how students' three-dimensional science learning builds over time with instruction. Beneficially, the process can be used to create tasks that are very different from traditional formats of assessment.

This book describes the details of our process and invites science educators to embark on a transformative journey to develop assessment tasks that will support them in their science instruction. An important aim is to shift the focus of classroom-based assessment away from serving primarily an evaluative purpose and to one that plays an instructionally supportive role. As you will find from this book, assessment can be valuable for classroom pedagogy, especially when it is integrated within instruction and used formatively to guide the progress of student learning. We encourage you to embrace the opportunity to stretch and reshape science assessment in your classroom.

With warm regards,
Christopher, Joe, and Jim

CHAPTER 1

The Critical Importance of Designing Assessment Tasks Aligned With the *NGSS*

James W. Pellegrino, University of Illinois Chicago • Joseph Krajcik,
Michigan State University • Christopher J. Harris, WestEd

An Inspiring Vision of Science Proficiency

We are experiencing an exciting and challenging time in science education. The National Research Council's (NRC) *A Framework for K–12 Science Education* (NRC, 2012) has put forth a vision for science proficiency that is quite bold and considerably different from previous conceptions of what students should know and be able to do in science (American Association for the Advancement of Science [AAAS, 1993]; NRC, 1996). In today's science classrooms, building science proficiency is not merely a matter of acquiring knowledge from a teacher, textbook, or video and being able to reproduce it. Rather, science proficiency requires a high degree of performance capability, whereby a learner must use and apply knowledge in varied and demanding ways. This proficiency develops over time and becomes stronger when a learner has opportunities to put knowledge to use. The vision for science proficiency that is the centerpiece of *A Framework for K–12 Science Education* (*Framework*) emphasizes providing solutions to complex problems and using and applying scientific knowledge to make sense of compelling phenomena.

The *Framework* describes three interconnected dimensions of scientific knowledge: disciplinary core ideas (DCIs), the big ideas associated with a discipline, like evolution from life science, which are essential to explain phenomena; crosscutting concepts (CCCs), ideas like systems thinking that are important across many science disciplines and provide

a unique lens to examine phenomena; and science and engineering practices (SEPs), the multiple ways of knowing and doing, like developing models and constructing explanations that scientists and engineers use to study the natural and designed world. The *Framework* focuses on the need for the integration of these three dimensions in science and engineering education. The knowledge associated with each of the three dimensions must be integrated in the teaching, learning, and doing of science and engineering, and in assessing what students know and can do. When learners engage in science and engineering practices that are integrated with DCIs and CCCs to make sense of compelling phenomena or design solutions to complex problems, they build new knowledge about all three dimensions and come to understand the nature of how scientific knowledge develops.

While each of the three dimensions matters, a central argument of the *Framework* is that proficiency is demonstrated through performances that require the integration of all three dimensions. Such performances have come to be known as *performance expectations* (PEs) because they specify what students at various levels of educational experience should know and be able to do. The *Next Generation Science Standards* (*NGSS*; NGSS Lead States, 2013) are an expression of the integrated knowledge vision contained in the *Framework*, and they provide a set of standards expressed as performances for students from kindergarten to 12th grade. The *NGSS* performance expectations move well beyond the vague terms typically used in previous science standards, such as *know* and *understand*, to more specific statements like *analyze and interpret*, *explain*, or *model*, in which the practices of science and engineering are wrapped around and integrated with disciplinary core ideas and crosscutting concepts. Thus, a student demonstrates grade-level proficiency by completing performances that show how they make use of knowledge. To truly know and understand science is to be able to use the three dimensions of scientific knowledge to explain compelling phenomena and provide solutions to complex problems.

The *Why* and the *What* Guiding This Book

Imagine a science teacher planning instruction to support students in acquiring the knowledge and capabilities underlying this performance expectation: *Analyze and interpret data on the properties of substances before and after the substances interact to determine if a chemical reaction has occurred.* This middle school physical science PE from the *NGSS* is a very challenging standard about what middle school students are expected to know and be able to do with their knowledge, and it is very different from previous science standards adopted by state departments of education. Prior standards typically focused on the science content alone or gave separate attention to content and inquiry. In those instances where states had both content and inquiry standards, the primary aim was to use inquiry as the means to learn the science content. Quite differently, performance expectations emphasize that it is not solely what students know, but also how students use and apply what they know that matters for making sense of the natural world and advancing in science learning.

Teachers throughout our nation are facing the challenge of preparing students to meet the requirements of such ambitious standards. Since the release of the *NGSS* in 2013, 49 states and the District of Columbia have adopted the *NGSS* or developed their own standards based on the *Framework*, representing a substantial proportion of the U.S. student population that is now expected to develop proficiency in performance standards (see https://www.nsta.org/science-standards). It is no surprise that considerable materials have been published in the fields of educational research and educational practice about the *Framework* and the *NGSS*, including NSTA Press books (e.g., Duncan et al., 2017; Nordine & Lee, 2021; Schwarz et al., 2017). These resources are designed to support teachers in planning their instruction, organizing their curricula, and understanding the expectations for students and teachers outlined in the *Framework* and the *NGSS*.

But what about assessment, especially classroom assessment designed to help teachers monitor and support the progress of their students' learning? Without good assessment tools, teachers will have a difficult time judging whether their students are making the progress needed to meet the *NGSS* performance expectations for their grade level or in their grade band. We know from considerable published literature and the wisdom of practice that assessment can be a valuable part of instructional practice, especially when integrated with curricular materials and lesson plans and used formatively to guide the progress of student learning (e.g., Penuel & Shepard, 2016). But we also know that these ambitious science standards pose considerable challenges when it comes to designing valuable assessments that support instructional practice and students' learning (NRC, 2014).

Imagine once again that a teacher is working to help students master the knowledge and capabilities required to meet the middle school physical science PE mentioned earlier. Figure 1.1 shows a task (*Miranda's Mystery Liquids*) that could be used to assess students' knowledge and capabilities for uncovering patterns through data analysis and interpretation on the properties of substances that are associated with chemical identity. The task requires students to determine which substances are the same or different based on patterns in data. A teacher can use this task to gauge students' understanding of which properties are associated with chemical identity, as well as students' ability to use and apply knowledge about the properties of matter. It can help a teacher determine if students are on the path toward meeting some of the key demands of the performance expectation.

FIGURE 1.1. Example assessment task: Miranda's Mystery Liquids

Miranda was responsible for cleaning up her work area and putting the materials away safely. During class, she used three different liquid substances, but after class, she found four unlabeled bottles of liquid by her desk. To put them away safely, she needs to know which liquids are the same and which are different.

To figure this out, Miranda measured the volume and mass of the liquids, which she used to calculate the density of each. She then tested the boiling point of the liquids. Table 1 shows the data from her investigation.

Table 1. Data of four liquids in different bottles.

Sample	Boiling Point	Mass	Volume	Density
1	100°C	6.10 g	6.10 cm³	1.00 g/cm³
2	126°C	5.39 g	6.10 cm³	0.883 g/cm³
3	78.4°C	8.05 g	10.2 cm³	0.789 g/cm³
4	126°C	9.01 g	10.2 cm³	0.883 g/cm³

A) Which information in the table would you use to tell Miranda whether any liquids could be the same substance? Be sure to tell why.

B) Based on the information in the table, which, if any, of the liquids are the same? Support your answer with what you know about the properties of matter.

This book focuses on how to approach and solve the problem of designing instructionally supportive assessments, such as the example provided above, that align with the science standards and are useful and usable in the classroom. The process discussed in this book will support teachers, instructional leaders and coaches, professional learning facilitators, state and district educational leaders, and curriculum and assessment developers who want to know how they can develop the types of tasks and problems that are faithful to the vision of integrated knowledge and practice as described in the *Framework* and the *NGSS*. To accomplish this, the book provides step-by-step guidance on how to use our innovative process, called the *Next Generation Science Assessment* (*NGSA*) design process, to create classroom-based assessment tasks and rubrics, and then illustrates how the tasks and rubrics can be used to support three-dimensional instruction. The process is a systematic, multi-step approach to designing assessment tasks that integrates disciplinary core ideas, science and engineering practices, and crosscutting concepts as called for in the *Framework* and the *NGSS*. Using this process will enable you to develop a variety of tasks that fulfill the important requirements for assessment of three-dimensional learning.

In the remainder of this chapter, we will review some ideas and principles associated with the *Framework* and the *NGSS* and their implications for instruction and assessment.

We will then review what it means for an assessment to be aligned to the *Framework* and the *NGSS* and practical in classroom instruction. Finally, we will preview the contents of the rest of this book, including descriptions of a systematic design process for developing equitable and fair assessments and advice regarding applications of those assessments in ongoing classroom practice.

The *Framework* and the *NGSS*: Science Education Visions and Challenges

Partly in reaction to criticisms of U.S. science curricula being "a mile wide and an inch deep" (Schmidt et al., 1997, p. 62) relative to other countries, the *Framework* identifies a small set of core ideas in four disciplines: (1) life sciences, (2) physical sciences, (3) Earth and space sciences, and (4) engineering, technology, and the application of science. The disciplinary core ideas (DCIs) are the big ideas associated with a discipline that are needed to make sense of phenomena and design solutions to problems encountered in the world. In putting forth a smaller set of core ideas, the *Framework* reduced the long and often disconnected catalog of factual knowledge that students previously had to learn. Disciplinary core ideas in the physical sciences include Energy and Matter, for example, and disciplinary core ideas in the life sciences include Ecosystems and Biological Evolution. Students are supposed to encounter these disciplinary core ideas over the course of their school years at increasing levels of sophistication, deepening their knowledge over time (see Duncan et al., 2017 for further details about disciplinary core ideas).

The second dimension described in the *Framework* is crosscutting concepts (CCCs). The *Framework* identifies seven such concepts that have importance across many science disciplines and are critical for making sense of phenomena and designing solutions to problems. Examples of crosscutting concepts include Patterns, Cause and Effect, Systems Thinking, and Stability and Change (see Nordine & Lee, 2021 for further information regarding crosscutting concepts).

The third dimension from this inspiring vision of science education is science and engineering practices (SEPs) that represent the multiple ways of knowing and doing that scientists and engineers use to study the natural and designed world. Eight key practices are identified, including Asking Questions (for science) and Defining Problems (for engineering), Planning and Carrying Out Investigations, Developing and Using Models, Analyzing and Interpreting Data, and Engaging in Argument from Evidence (see Schwarz et al., 2017 for further information about science and engineering practices).

The disciplinary core ideas, crosscutting concepts, and science and engineering practices serve as thinking tools that work together to enable scientists, engineers, and learners to design solutions to problems, reason with evidence, and make sense of phenomena.

The *Framework* makes the case that proficiency and expertise develop over time and increase in sophistication as the result of learners experiencing coherent, integrated systems of curriculum, instruction, and assessment. To illustrate the instructional and assessment challenges posed by the vision of proficiency in the *Framework*, consider the projected end point of K–12 science education. By the end of 12th grade, all students—not just those interested in pursuing science, engineering, or technology studies beyond high school—should have gained sufficient knowledge and understanding to:

1. engage in public discussions of science-related issues such as the challenges of generating sufficient energy, preventing and treating diseases, maintaining supplies of clean water and food, and addressing problems caused by global environmental change;

2. be critical consumers of scientific information related to their everyday lives; and

3. continue to learn about science throughout their lives.

Students should come to appreciate that science as a discipline and the current scientific understanding of the world are the result of hundreds of years of creative human endeavor (NRC, 2012, p. 24).

The *NGSS* use the three dimensions to organize the content and sequence of learning in a way designed to meet the ambitious goals outlined in the *Framework*. The three-part structure of a performance expectation that integrates a DCI, a CCC, and an SEP signals an important revolutionary shift for science education and presents the primary challenge for the design of both instruction and assessment: finding a way to describe and capture students' developing proficiency along these intertwined dimensions (see Figure 1.2). Research shows that developing knowledge-in-use involves simultaneously using DCIs, CCCs, and SEPs to make sense of phenomena and find solutions to complex problems (National Academies of Sciences Engineering and Medicine, 2019; NRC, 2012). As mentioned above, science and engineering practices, crosscutting concepts, and disciplinary core ideas are envisaged as tools for making sense of phenomena and designing solutions to problems. Students who experience the use of these dimensions in multiple contexts are more likely to become flexible and effective users of all three forms of scientific knowledge in new problem contexts.

FIGURE 1.2. Intertwined dimensions of science proficiency

However, disciplinary core ideas should not be equated with traditional content ideas and science and engineering practices should not be equated with inquiry skills. Instead, disciplinary core ideas are the big ideas within a discipline that explain a host of phenomena, and science and engineering practices provide learners with the means to engage in making sense of phenomena. Although science and engineering practices have some overlap with various inquiry skills, the use of science and engineering practices is intended to emphasize both knowledge and capability. The knowledge includes development of an understanding of each practice—what it is and why it is important. For instance, what constitutes a model, why models change, and why constructing and using models are important in science. Capability reflects knowledge of how to carry out specific practices like how to develop a model. It is critical to realize that disciplinary core ideas and science and engineering practices, as well as crosscutting concepts, develop over time as students use them to make sense of phenomena and design solutions to problems.

The *Framework* uses learning progressions, statements about how knowledge develops over time in a coherent fashion, to describe students' developing proficiency in each of these three intertwined dimensions across grades K–12. Well-designed learning progressions provide a map of the routes that can be taken by students from the earliest grades onward as they progress across the grade levels (NRC, 2012). The emphasis on the developmental nature of learning and learning progressions is supported by research on learning (see for instance Alonzo & Gotwals, 2012; Corcoran et al., 2009; NRC, 2007).

In the context of assessment, the importance of this integrated perspective of what it means to know and be able to do science is that a disciplinary core idea, crosscutting concept, or science and engineering practice should become successively more sophisticated over time. This level of sophistication needs to be assessed. Assessments need to be written so that the level of students' understanding can be tracked from less sophisticated to more sophisticated to ensure students are making progress toward the performance expectation. This is a relatively unfamiliar idea in the realm of science assessments, which have more often been viewed as simply measuring whether or not students know particular grade-level content. The concept of tracking a level of understanding means that assessments must strive to be sensitive both to grade-level-appropriate understanding and to those understandings that may be appropriate at somewhat lower or higher grades. This is particularly important for assessment materials and resources that can be used to support classroom instruction. Take, for example, the following performance: *Students construct a model to show why a rolling car will eventually stop or might change direction.* Third graders can use the idea of balanced and unbalanced forces to draw such a model. These ideas can also be assessed at the middle and high school level, but the level of sophistication will be much greater in subsequent years. The foundation of developing an understanding of force occurs early in the elementary grades, and from this foundation more sophisticated understanding of this challenging idea can be built.

To support such an integrated approach to science learning, the *Framework* explains that "Assessment tasks must be designed to gather evidence of students' ability to apply the practices and their understanding of the crosscutting concepts in the contexts of problems that also require them to draw on their understanding of specific disciplinary ideas" (NRC, 2012). In developing the *NGSS* (NGSS Lead States, 2013), the *NGSS* committee generated standards for each grade at the K–5 level and at the middle- and high-school grade bands in which learners are expected to apply the practices, crosscutting concepts, and disciplinary knowledge they develop.

At the start of this chapter, we provided an example of one performance expectation for middle school physical science. Often instructional designers of curriculum materials and assessments will cluster or bundle together performance expectations related to a particular aspect of a disciplinary core idea for coherence and appropriate classroom use. Figure 1.3 provides an example of such a bundle of performance expectations for fourth grade life science with details related to the three-dimensional architecture of this set of standards. Each of the performance expectations shown asks students to use a specific practice and a crosscutting concept in the context of a specific element of the disciplinary knowledge relevant to the particular aspect of the core idea. Across performance expectations at a given grade level or grade band, each practice and crosscutting concept appears multiple times.

FIGURE 1.3. Bundle of performance expectations for fourth grade life science

Students who demonstrate understanding can:
- **4-LS1-1. Construct an argument that plants and animals have internal and external structures that function to support survival, growth, behavior, and reproduction.** [Clarification Statement: Examples of structures could include thorns, stems, roots, colored petals, heart, stomach, lung, brain, and skin.] [Assessment Boundary: Assessment is limited to macroscopic structures within plant and animal systems.]
- **4-LS1-2. Use a model to describe that animals receive different types of information through their senses, process the information in their brain, and respond to the information in different ways.** [Clarification Statement: Emphasis is on systems of information transfer.] [Assessment Boundary: Assessment does not include the mechanisms by which the brain stores and recalls information or the mechanisms of how sensory receptors function.]

The performance expectation above was developed using the following elements from the NRC document *A Framework for K–12 Science Education*:

Science and Engineering Practices	Disciplinary Core Ideas	Crosscutting Concepts
Developing and Using Models Modeling in 3–5 builds on K–2 experiences and progresses to building and revising simple models and using models to represent events and design solutions. - Use a model to test interactions concerning the functioning of a natural system. (4-LS1-2) **Engaging in Argument from Evidence** Engaging in argument from evidence in 3–5 builds on K–2 experiences and progresses to critiquing the scientific explanations or solutions proposed by peers by citing relevant evidence about the natural and designed world(s). - Use a model to test interactions concerning the functioning of a natural system. (4-LS1-2) - Construct an argument with evidence, data, and/or a model. (4-LS1-1)	**LS1.A: Structure and Function** - Plants and animals have both internal and external structures that serve various functions in growth, survival, behavior, and reproduction. (4-LS1-1) **LS1.D: Information Processing** - Different sense receptors are specialized for particular kinds of information, which may be then processed by the animal's brain. Animals are able to use their perceptions and memories to guide their actions. (4-LS1-2)	**Systems and System Models** - A system can be described in terms of its components and their interactions. (4-LS1-1), (4-LS1-2)

Connections to other DCIs in this grade band: N/A

Articulation of DCIs across grade bands:
1.LS1.A (4-LS1-1); 1.LS1.D (4-LS1-2); 3.LS3.B (4-LS1-1); MS.LS1.A (4-LS1-1), (4-LS1-2); MS.LS1.D (4-LS1-2)

Common Core State Standards Connections:

ELA/Literacy

W.4.1	Write opinion pieces on topics or texts, supporting a point of view with reasons and information. (4-LS1-1)
SL.4.5	Add audio recordings and visual displays to presentations when appropriate to enhance the development of main ideas or themes. (4-LS1-2)

Mathematics

4.G.A.3	Recognize a line of symmetry for a two-dimensional figure as a line across the figure such that the figure can be folded across the line into matching parts. Identify line symmetric figures and draw lines of symmetry. (4-LS1-1)

As shown in Figure 1.3, the *NGSS* performance expectations reflect integrations of elements of the disciplinary core ideas, science and engineering practices, and crosscutting concepts. Because DCIs, SEPs, and CCCs are large in scope, they are broken down into elements in the *NGSS* performance expectations. The dimension boxes under the performance expectations identify the elements of the dimensions used in the performance expectation that are most appropriate for the development of learners in that grade band or grade level. You can identify the elements for each dimension by looking for the elements in the dimension boxes.[1] Performance expectations also include boundary statements that identify limits to the level of understanding or context appropriate for a particular grade level and clarification statements that offer additional details and examples.

Although the performance expectations provide a large degree of specificity, even as explicated in the *NGSS,* they do not provide sufficient detail to create usable assessment tasks. In fact, because the performance expectations are at such a large grain size, to develop valid, reliable, and useful science assessment tasks that align with the performance expectations, such as those in Figure 1.3, assessment designers need to take into consideration many features: (1) the various sub-ideas contained in the performance expectations; (2) smaller-scale learning goals that focus on performances (we refer to these as *learning performances*, which are knowledge-in-use statements that incorporate aspects of disciplinary core ideas, science and engineering practices, and crosscutting concepts that students need to develop an understanding of as they progress toward achieving PEs); (3) grade-level-appropriate and motivating scenarios for assessing the learning performances; (4) the types of evidence that will reveal levels of student understanding and capabilities; (5) options for task design features (e.g., more or less scaffolding); and (6) equity and inclusion considerations so that all learners can understand and relate to the assessment tasks. We elaborate on these ideas in subsequent chapters.

Instructionally Supportive Classroom Assessment

The *Framework* and the *NGSS* performance expectations raise many questions about what valid science assessments should look like and how to design them for various purposes, including use in the classroom. For example, teachers use various forms of assessment to inform day-to-day and month-to-month decisions about instruction and learning, and to track student progress and give students feedback on their progress. Such assessments, commonly referred to as the *formative use of assessment* (see Black & Wiliam, 1998;

---

1   For further information see Appendixes F and G of the *NGSS.*

Wiliam, 2007), provide specific information about students' strengths and difficulties with learning. Teachers can use this information to adapt their instruction to meet student needs, which likely will vary from one student to another. Students can also use this information to determine whether they need to focus on specific types of capabilities or knowledge or make certain adjustments to their thinking.

Teachers and school districts also conduct assessments to help determine whether a student has attained a certain level of proficiency after completing a particular phase of education, whether a two-week curriculum unit, a semester-long course, or one or more years of schooling. This is referred to as an *assessment of individual achievement*, or the *summative use of assessment*. Some of the most familiar forms of summative assessment are those used by classroom teachers, such as end-of-unit or end-of-course tests, which often are used to assign letter grades when a course is finished. Large-scale assessments—which are administered at the direction of external users like a district or state—also provide information about the achievement of individual students, as well as comparative information about how one individual performed relative to others. Because large-scale assessments are typically given only once a year and involve a significant time lag between testing and availability of results, the results seldom provide information that can be used to help teachers or students make day-to-day or month-to-month decisions about teaching and learning.

In this book, we focus on the design of assessments for classroom use with an emphasis on *assessment for learning*. This is a challenge given the very nature of the *NGSS* performance expectations, as they are large in scope and describe what students should know and be able to do at the end of each grade level for K–5 and at the end of each grade band for middle school and high school. We ask: *How can teachers support student learning and assess learning directed toward such ambitious goals?* We tackle this problem by breaking down these complex performance expectations into smaller, three-dimensional performance statements that can be used to design instructionally supportive assessment tasks to provide meaningful feedback to students and teachers. We refer to these smaller targets as learning performances (LPs). For instance, the performance expectation MS-LS1-6 can be broken down into several learning performances.

Performance Expectation:

- MS-LS1-6. Construct a scientific explanation based on evidence for the role of photosynthesis in the cycling of matter and flow of energy into and out of organisms.

Example Learning Performances:

- LP 1: Students analyze and interpret data to determine whether plants and other photosynthetic organisms grow with the input of energy from sunlight.
- LP 2: Students analyze and interpret data to determine whether plants and other photosynthetic organisms take in water, carbon dioxide, and energy (e.g., sunlight) to make food (sugar) and oxygen.
- LP 3: Students develop a model that shows that plants (or other photosynthetic organisms) take in water and carbon dioxide to make food (sugar) and oxygen.

Notice how each learning performance includes a science and engineering practice, elements of the disciplinary core idea, and a crosscutting concept. The set of learning performances provides information on how students develop proficiency toward the performance expectation. In Chapter 4, we present a detailed look at the structure of learning performances and describe how to develop them so that they can help guide the design of instructionally supportive classroom assessments.

As the name implies, instructionally supportive assessments should provide actionable information of value to teachers and their students. Unfortunately, much of the discussion regarding the design and use of educational assessment has tended to focus on large-scale standardized tests, especially given their prominent use in the United States for purposes of school and teacher accountability. Such assessments have limited instructional utility given the nature of their design. Further complicating matters is the fact that teachers tend to emulate such assessments when they develop assessments for use in their own classrooms. Only within the last 15–20 years has there been a significant effort to acknowledge that assessments intended to function close to classroom teaching and learning require greater attention with regard to processes of design and validation, given their importance for supporting the attainment of significant educational goals such as the vision for science learning found in the *Framework* and the *NGSS* (see e.g., Armour-Thomas & Gordon, 2013; NRC, 2014; Pellegrino et al., 2016).

A major premise of the National Research Council's report *Knowing What Students Know: The Science and Design of Educational Assessment* (Pellegrino et al., 2001) is that the effectiveness and utility of educational assessments must ultimately be judged by the extent to which they help promote student learning. The aim of assessment should be "*to educate and improve* student performance, not merely to *audit* it" (Wiggins, 1998, p. 7).

How to Design and Use Instructionally Supportive Science Assessments

Because of the challenge to develop assessments that align to *Framework*-inspired science standards, it should come as no surprise that comprehensive sets of examples of the types of assessments that align with the performance expectations in these standards are lacking. Many of the science assessment tasks that have typically been used for classroom assessment, as well as those found in large-scale state, national, and international tests, focus primarily on science content or on aspects of scientific inquiry separate from content. With relatively few exceptions, such assessments do not integrate disciplinary core ideas, crosscutting concepts, and science and engineering practices in the ways intended by the *Framework*.

Fortunately, some of what we now know about the design of educational assessments can be productively used to develop science assessments that represent the types of tasks and situations called for by the *Framework* and the *NGSS*. We can draw approximations from cases that illustrate different forms of science assessment. The NRC report *Developing Assessments for the Next Generation Science Standards* (NRC, 2014) describes several of these cases, which include topics related to biodiversity in the schoolyard and climate change. These examples, along with other cases, are diverse in terms of science content and practices represented, age and grade level, and scale or use (i.e., at the classroom, state, or national level). They are also delivered using different methods (some with technology and some without) and include different consequences of student performance (some with high stakes and some with low stakes).

A very important point that was emphasized in *Developing Assessments for the Next Generation Science Standards* (NRC, 2014) is the need for a principled approach to assessment development, especially given the ambitious nature of the standards and their complexity in terms of the knowledge and capabilities that students are expected to demonstrate. Unfortunately, the tendency in assessment development has often been to work from a somewhat "loose" description of what students are supposed to know and be able to do and move directly to the development of tasks or problems for them to answer. Such a description might be a general statement about what the assessment needs to cover and some indication about the types of tasks to use (e.g., multiple choice, short constructed response) rather than the forms of evidence to be derived from the specific tasks given to a student and what that evidence would mean. It is then left up to the item writer to generate some candidate items. Given the complexities of the assessment design process, it is unlikely that such a diffuse process could lead to the generation of quality

assessment tasks without a great deal of artistry, luck, trial and error, and maybe a miracle or two. Figure 1.4 shows a well-known cartoon by Harris that captures the essence of the problem—much of typical assessment development relies on "miracles" in going from standards to assessments when in fact assessment designers need to be much more explicit about how to move from standards to assessments. Without a principled approach, many assessments will not cover all of the dimensions and there may be uncertainty about the conclusions that can be drawn from student performance on the various tasks.

FIGURE 1.4. Cartoon depicting the need to explain each part of a process

Reprinted with permission from S. Harris, sciencecartoonsplus.com

Our argument is that assessment design for the *NGSS* is unlikely to succeed without explicit attention to what should be assessed and how to systematically go about designing tasks to obtain the information needed to understand what students actually know and can do. This knowledge has implications for both instruction and understanding levels of student learning. In the remainder of this book, we present a systematic design approach to develop such assessments. They are intended for classroom use and are thoughtfully aligned to the *Framework* and the *NGSS*. To develop such assessments, we do not want to rely on miracles—instead, we will rely on a systematic process to go from the standards to tasks that teachers can use in classrooms to inform teaching and learning. One aspect of the "validity argument" for instructionally supportive assessments is a clear articulation of the design process itself (see Pellegrino et al., 2016). We claim that, if you follow the process outlined in the remaining chapters of this book, you can make substantial progress in creating valid, usable, and instructionally useful assessments of students' knowledge and capabilities associated with the *NGSS* performance expectations.

Overview of the Book

In this section, we present a short overview of this book and where it will take you—and the benefits if you go down this path. Our goal is that, by reading and working through the various chapters in the book, you will develop the knowledge and capabilities to design assessments that align to the *NGSS* performance expectations or similar types of state standards. Each chapter systematically builds on the previous chapters.

In Chapter 2, ***Creating Assessment Tasks for NGSS Classrooms: An Overview of the Design Process***, we present an introduction to the *NGSA* process that we use to design assessment tasks and describe its value for meeting the important requirements for assessment of three-dimensional learning. The chapter is guided by the following question: *What approach can be used to create assessments that will help you and other teachers know whether instructional experiences make a difference for students in building their three-dimensional learning?*

Chapter 3, ***Unpacking and Mapping the* NGSS *Dimensions***, explores selecting performance expectations and takes a deep dive into how to unpack the meaning of each dimension in a performance expectation or bundle of performance expectations. The following question is explored: *How can you and other teachers deconstruct and make sense of performance expectations to accurately determine what students should know and be able to do?* After we provide a description of the unpacking process, we present a technique for bringing key aspects of the unpacked dimensions back together in a visual representation that we call integrated dimension maps.

Chapter 4, ***Constructing Learning Performances That Build Toward the* NGSS *Performance Expectations***, presents one of the unique aspects of our process for designing assessment tasks that build proficiency toward achieving the performance expectations. In this chapter, we introduce the idea of *learning performances* as three-dimensional learning goals that are smaller in grain size than the performance expectations. They take the form of knowledge-in-use statements that incorporate aspects of DCIs, SEPs, and CCCs that students need to develop understanding of as they progress toward achieving a single performance expectation or bundle. Chapter 4 responds to the following question: *How can you and other teachers use performance expectations to construct assessment tasks that can be used during instruction?*

In Chapter 5, ***Developing Assessment Tasks That Provide Evidence of Three-Dimensional Learning***, we introduce design blueprints as an all-important organizer for using learning performances to construct assessment tasks that assess for three-dimensional learning. The chapter explores the following question: *How can assessment tasks be designed to provide evidence of three-dimensional learning so that you and other teachers can gauge students' progress with the* NGSS *performance expectations?* We begin by describing what design blueprints are and how they provide guidance for task design. Then, we detail how to specify design blueprints and use them to construct assessment tasks that align with learning performances. The chapter includes guidance for selecting phenomena and creating motivating scenarios for assessment tasks.

Chapter 6, ***Attending to Equity and Inclusion in the Assessment Design Process***, presents design principles and considerations for ensuring that tasks are accessible and fair for a wide range of students with varying backgrounds, skills, and abilities so that they can demonstrate three-dimensional learning. The foremost question guiding the chapter asks: *How can you and other teachers create tasks that will both leverage and value students' background knowledge and experiences and connect these to rigorous science learning?* The chapter concludes with discussion of some of the unique ways in which technology can be used in the design and delivery of assessment tasks to support equity and inclusion.

In Chapter 7, ***Developing and Using Rubrics That Integrate the Three Dimensions of Science Proficiency***, we provide guidance on how to create rubrics for tasks in which learners need to use the three dimensions to make sense of phenomena or solve complex problems. We focus our attention on rubrics as tools for teaching and for promoting student learning. We also offer practical guidance for how to use rubrics to gain insight into students' performance and use the information to improve teaching and advance learning through actionable feedback. The chapter provides guidance for answering the question: *How can rubrics be developed and used that will help you and other teachers to "see" students' progress in building three-dimensional proficiency?*

Chapter 8, ***Using Assessment Tasks to Promote Student Learning***, describes how teachers can use three-dimensional assessment tasks in the midst of classroom instruction, including ways to incorporate assessment tasks into classroom activities. The chapter explores the question: *How can you and other teachers use assessment tasks during instruction to learn about students' performance and support their three-dimensional learning?* In response to this question, we offer practical guidance for how teachers can take steps toward using tasks in innovative ways that stretch far beyond traditional uses for classroom assessment.

Finally, Chapter 9, ***Reflections and Implications: Creating and Using Three-Dimensional Assessment Tasks to Support NGSS Instruction***, discusses the benefits and challenges when it comes to designing high-quality assessments that support instructional practice and students' learning. The chapter includes a reflection on the *NGSA* design process and its value for creating tasks consistent with the *Framework* and the *NGSS*. We conclude the book with implications for design and classroom use of tasks and consider the emerging role of technology for empowering teachers and students to use assessments in new and innovative ways.

References

American Association for the Advancement of Science (AAAS). 1993. *Benchmarks for science literacy*. New York: Oxford University Press.

Alonzo, A. C. & A. W. Gotwals, eds. 2012. *Learning progression in science: Current challenges and future directions*. Rotterdam, Netherlands: Sense.

Armour-Thomas, E. & E. W. Gordon. 2013. *Toward an understanding of assessment as a dynamic component of pedagogy*. Princeton NJ: Gordon Commission.

Black, P. & D. William. 1998. Assessment and classroom learning. *Assessment in Education: Principles, Policy & Practice,* 5(1), 7–73.

Corcoran, T. B., F. A. Mosher & A. Rogat. 2009. *Learning progressions in science: An evidence-based approach to reform.* New York, NY: Columbia University, Teachers College, Consortium for Policy Research in Education, Center on Continuous Instructional Improvement.

Duncan, R. G., J. Krajcik & A. E. Rivet, eds. 2016. *Disciplinary core ideas: Reshaping teaching and learning*. Arlington, VA: National Science Teachers Association Press.

National Academies of Sciences Engineering and Medicine. 2019. *Science and engineering for grades 6–12: Investigation and design at the center*. Washington, DC: National Academies Press.

National Research Council (NRC). 2014. *Developing assessments for the Next Generation Science Standards*. Washington, DC: National Academies Press.

National Research Council (NRC). 2012. *A framework for K–12 science education*. Washington, DC: National Academies Press.

National Research Council (NRC). 1996. *National Science Education Standards*. Washington, DC: National Academy Press.

National Research Council (NRC). 2007. *Taking science to school: Learning and teaching science in grades K–8*. Washington, DC: National Academies Press.

NGSS Lead States. 2013. *Next Generation Science Standards: For states, by states*. Washington, DC: National Academies Press.

Nordine, J. & O. Lee, eds. 2021. *Crosscutting concepts: Strengthening science and engineering and learning.* Arlington, VA: National Science Teachers Association.

Pellegrino, J. W., N. Chudowsky, & R. Glaser (Eds.). (2001). *Knowing what students know: The science and design of educational assessment.* National Academies Press.

Pellegrino, J. W., L. V. DiBello & S. R. Goldman. 2016. A framework for conceptualizing and evaluating the validity of instructionally relevant assessments. *Educational Psychologist*, 51(1), 59–81.

Penuel, W. R. & L. A. Shepard. 2016. Assessment and teaching. In *Handbook of Research on Teaching, 5th edition*, ed. D. H. Gitomer & C. A. Bell, 787–850. Washington, DC: American Educational Research Association.

Schmidt, W. H., C. C. McKnight & S. A. Raizen. 1997. *A splintered vision: An investigation of U.S. science and mathematics education.* Boston, MA: Kluwer Academic.

Schwarz, C. V., C. Passmore & B. J. Reiser. 2017. *Helping students make sense of the world using next generation science and engineering practices.* Arlington, VA: National Science Teachers Association Press.

Wiggins, G. 1998. *Educative assessment: Designing assessments to inform and improve student performance.* San Francisco: Jossey-Bass.

Wiliam, D. 2007. Keeping learning on track: Formative assessment and the regulation of learning. In *Second handbook of mathematics teaching and learning*, ed. F. K. Lester, Jr., 1053–1098. Greenwich, CT: Information Age.

CHAPTER 2

Creating Assessment Tasks for *NGSS* Classrooms: An Overview of the Design Process

Brian D. Gane, University of Kansas • Samuel Severance, Northern Arizona University • Joseph Krajcik, Michigan State University

The *Framework for K–12 Science Education* (*Framework;* National Research Council, 2012) and *Next Generation Science Standards* (*NGSS;* NGSS Lead States, 2013) require broad shifts in how we conceptualize science learning and how we assess students' progress. Perhaps the most noteworthy shift is from a view of science learning as primarily a process of acquiring science content knowledge toward a view of learning as a process of using and applying disciplinary core ideas in concert with science and engineering practices and crosscutting concepts to make sense of phenomena or solve real world problems (National Research Council, 2012). Referred to as three-dimensional learning, this view emphasizes using and applying the three dimensions in an integrated manner as the means for building the proficiencies expected by the *NGSS* Performance Expectations (PEs). To support and sustain teachers and students in this type of learning, assessment tasks for *NGSS* classrooms must also be three-dimensional so that they can provide information about how students are progressing toward achieving the PEs (National Research Council, 2014). This change in assessment practice represents another important shift for science education. As underscored in Chapter 1, this shift is a different way of thinking about assessment where what matters for measuring students' performance is not just what students know, but also how they use and apply what they know. In order to ensure that we measure what matters for science instruction, we need to begin with a foundational question: *What approach can be used to create assessments that will help you and other teachers know whether instructional experiences make a difference for students in building their three-dimensional learning?*

In this chapter, we introduce the *Next Generation Science Assessment (NGSA)* design process. The process is a systematic multi-step approach for designing classroom-based assessment tasks that provide evidence of three-dimensional learning so that teachers can gauge students' progress with the *NGSS* PEs. As you will learn, the process reflects an accessible vision for how to design three-dimensional assessment tasks. It also provides a structure that will add consistency for developing a variety of tasks that share common design elements and that align with their target PEs. Using this approach will enable you to fulfill the important requirements for assessment of three-dimensional learning.

We begin by casting a critical eye on an example assessment task and use this task as a starting point for considering what is needed for assessment tasks in general to be instructionally supportive for *NGSS* classrooms. Then, we provide an overview of the *NGSA* design process, briefly describing its six major steps for moving from a PE or bundle of PEs to a set of three-dimensional tasks for classroom use. We end the chapter by summarizing the primary takeaways of using this systematic approach for creating assessment tasks that align with PEs and that provide actionable information about how students are progressing toward meeting them.

Critical Examination of a Three-Dimensional Assessment Task

Imagine an assessment task that was created to align with the three-dimensional view of science learning and could be used to judge students' progress toward meeting an *NGSS* performance expectation. *What might this assessment task look like? How could science teachers use this assessment task in their classrooms to get a sense of what their students know and can do? Would an assessment task that measures three-dimensional learning need to fully address all components of a PE?* Figure 2.1 shows an example assessment task, *Miranda's Mystery Liquids*. The task relates to the physical science topic of Matter and its Interactions and aims to measure students' progress toward meeting MS-PS1-2, a PE from the middle school grade band (see Figure 2.2). *Does responding to this task provide information to the teacher and students as to whether students are building knowledge toward the PE?* In this task, students are presented with a brief scenario-based problem in which they need to use data analysis and interpretation to determine which, if any, of the unknown liquids in a collection might be the same substance. The task requires students to identify relevant patterns in the data table and apply knowledge about the characteristic properties of matter. Students also need to provide evidence and reasoning for what led them to make their determination.

Take a moment to read through the task and then respond to it.

FIGURE 2.1. Example assessment task: Miranda's Mystery Liquids

Miranda was responsible for cleaning up her work area and putting the materials away safely. During class, she used three different liquid substances, but after class, she found four unlabeled bottles of liquid by her desk. To put them away safely, she needs to know which liquids are the same and which are different.

To figure this out, Miranda measured the volume and mass of the liquids, which she used to calculate the density of each. She then tested the boiling point of the liquids. Table 1 shows the data from her investigation.

Table 1. Data of four liquids in different bottles.

Sample	Boiling Point	Mass	Volume	Density
1	100°C	6.10 g	6.10 cm³	1.00 g/cm³
2	126°C	5.39 g	6.10 cm³	0.883 g/cm³
3	78.4°C	8.05 g	10.2 cm³	0.789 g/cm³
4	126°C	9.01 g	10.2 cm³	0.883 g/cm³

A) Which information in the table would you use to tell Miranda whether any liquids could be the same substance? Be sure to tell why.
B) Based on the information in the table, which, if any, of the liquids are the same? Support your answer with what you know about the properties of matter.

After working through this task, consider how it relates to the PE from which it was developed, and how you would respond to the following questions:

1. To what extent does the task match with the dimensional elements of the PE?

2. What type of information can this task provide about students' proficiency with the PE?

3. How would this task be helpful in monitoring students' progress in building toward the PE?

FIGURE 2.2. Performance expectation MS-PS1-2 with foundation boxes (NGSS Lead States, 2013). Performance expectations integrate the three dimensions; the foundation boxes provide further information about each dimension.

MS-PS1-2. Analyze and interpret data on the properties of substances before and after the substances interact to determine if a chemical reaction has occurred. [Clarification Statement: Examples of reactions could include burning sugar or steel wool, fat reacting with sodium hydroxide, and mixing zinc with hydrogen chloride.] [*Assessment boundary: Assessment is limited to analysis of the following properties: density, melting point, boiling point, solubility, flammability, and odor.*]

The performance expectation above was developed using the following elements from the NRC document *A Framework for K–12 Science Education*:

Science and Engineering Practices	Disciplinary Core Ideas	Crosscutting Concepts
Analyzing and Interpreting Data • Analyzing data in 6–8 builds on K–5 and progresses to extending quantitative analysis to investigations, distinguishing between correlation and causation, and basic statistical techniques of data and error analysis. • Analyze and interpret data to determine similarities and differences in findings.	**PS1.A: Structure and Properties of Matter** • Each pure substance has characteristic physical and chemical properties (for any bulk quantity under given conditions) that can be used to identify it. **PS1.B: Chemical Reactions** • Substances react chemically in characteristic ways. In a chemical process, the atoms that make up the original substances are regrouped into different molecules, and these new substances have different properties from those of the reactants.	**Patterns** • Macroscopic patterns are related to the nature of microscopic and atomic-level structure.

When responding to the first question, you might have noticed a mismatch between the knowledge and capabilities that the assessment task asks students to demonstrate and what is fully required in the PE. For instance, *Miranda's Mystery Liquids* requires students to use only some elements of the disciplinary core idea (DCI). Students are only applying knowledge about the properties of matter in order to distinguish substances from one another. The task does not involve applying another important element of the DCI, which is focused on chemical reactions and the processes underlying them. Moreover, while both the PE and the task share the same science and engineering practice (SEP) of Analyzing and Interpreting Data, they differ somewhat in their application of the SEP. In *Miranda's Mystery Liquids*, the SEP is applied to determine similarities and differences between characteristic properties of substances. This is in contrast to the requirement of the PE, which is to determine similarities and differences before and after substances interact. The crosscutting concept (CCC) of Patterns is also the same for the task and the PE, but the task emphasizes patterns at the macroscopic scale whereas the PE places the emphasis on relating macroscopic patterns to the atomic level. Overall, the task matches with a number of important dimensional elements of the PE, but it does not fully cover all the PE's terrain.

The second and third questions might be a little harder to answer. *What type of information does this task stand to provide about students' proficiency with the PE? How would this task be helpful in monitoring students' progress in building toward the PE?* Because *Miranda's Mystery Liquids* addresses some of the key demands of the PE, the task at least stands to provide some valuable insight into students' developing proficiency. Noteworthy is that the task elicits some of the knowledge and capabilities students need to demonstrate in order to attain proficiency with the PE. For instance, being able to determine whether substances are the same or different based upon patterns in characteristic properties, as called for in the task, is a smaller three-dimensional performance that is needed for achieving the more comprehensive PE. Accordingly, it may not be a useful task for summative assessment purposes, but it holds promise for use during instruction at a time point when students would be expected to demonstrate proficiency with this essential part of the PE. The value of assessing students on a smaller three-dimensional performance like the one addressed in *Miranda's Mystery Liquids* is that it can help us to see whether students are on a path for building toward and successfully achieving the comprehensive PE. Tasks like these can provide teachers with "just-in-time" information that they can use to determine next steps for teaching and learning while in the midst of instruction. As will be described later in this chapter and elaborated on in Chapters 3 and 4, two features used in our *NGSA* design process—focusing on purposefully selected elements of the DCI, SEP, and CCC, and on occasion swapping in a different but closely related SEP or CCC—are done intentionally to create tasks that cohere and build with instruction over time. As part of this process, we break down PEs into comprehensive sets of smaller performances that in turn can be used to develop assessment tasks suitable for *NGSS* instruction. This enables us to create *instructionally supportive* tasks that can serve as markers to help teachers and students gauge how students are building in their proficiencies.

The Importance of Assessing Students' Progress Toward Achieving a PE

To understand how to assess a PE, one needs a deep understanding of all the elements encompassed within the PE. This is an important and worthwhile undertaking for anyone planning to develop three-dimensional tasks. If you have spent some time examining the *NGSS* PEs, you might have noticed several overarching characteristics:

- PEs are *statements* that describe what students should know and be able to do at the end of a grade or grade band. They are very broad in scope and were developed from the *Framework* (National Research Council, 2012). Each PE takes the form of a single statement that describes competencies at a large grain size, rather than going into the details that underlie them.
- PEs are *three-dimensional*: A PE always includes a DCI, SEP, and CCC. The dimensions are described in the *Framework*, and their elements (i.e., essential aspects of each dimension) are further described in the *NGSS*. The elements of the dimensions vary from PE to PE within a grade level or grade band, and they become more sophisticated as students progress through K–12.
- The three dimensions in the PEs are *integrated*: They describe how students use DCIs and CCCs by engaging in SEPs to make sense of phenomena or solve problems. The dimensions are not intended to be used in isolation from one another.

Expanding on these characteristics, a PE is analogous to an iceberg. The small portion of the iceberg that shows above the water's surface seems easy to navigate, but below the water's surface lies the remainder: a massive portion of the overall iceberg (see Figure 2.3). Similarly, a succinct PE statement such as MS-PS1-2, "Students who demonstrate understanding can analyze and interpret data on the properties of substances before and after the substances interact to determine if a chemical reaction has occurred," is really just the tip of the iceberg; students need to develop multiple proficiencies as they move toward achieving this PE. Just as it is unwise to ignore the portion of an iceberg below the water's surface when traveling through icy waters, we should not ignore the details that underlie each dimension within a PE. This becomes readily apparent when we consider the surface-level features of MS-PS1-2. For example, the DCI portion of the PE indicates that students will apply their knowledge about the properties of substances to determine if a chemical reaction has occurred. But what is expected of middle school students' knowledge of properties, substances, and chemical reactions? These details are what lies below the tip of the iceberg. The CCC portion, Patterns, is not explicit in the one-sentence PE statement, yet it is essential for students to identify patterns and reason with patterns in order to achieve this PE. *What are the types of patterns that students will need to use?* The SEP is clearly stated in the PE, with a focus on analysis of data to identify similarities and differences before and after substances interact. *But what grade level skills and abilities related to the SEP will students be expected to use?* By examining all that lies beneath the tip of the iceberg—or in this case the PE statement—we can more readily grasp the vast expanse. Once this expanse is mapped, it becomes clear that students' learning toward meeting a PE or a PE bundle must encompass a surprising amount.

FIGURE 2.3. Immediately visible portion of an *NGSS* Performance Expectation and what is called for beneath the surface

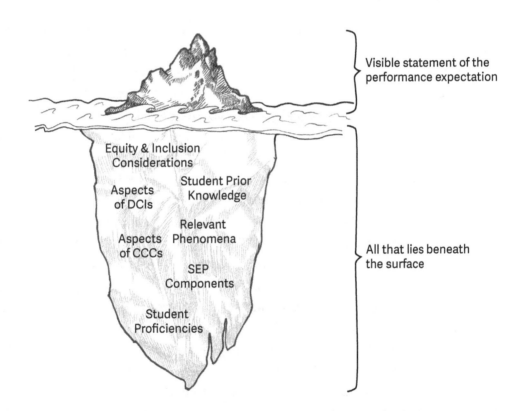

The *NGSA* design process is intended to assist task designers, whether classroom teachers, state and district educational leaders, or curriculum and assessment developers, in navigating through the PE icebergs with an eye toward understanding all of what lies beneath the surface. PEs are comprehensive and considered summative goals, and therefore need to be learned over time and through a sequence of lessons and instructional units. Alongside instruction, assessment tasks are needed that can be used—over time

and in a formative manner—to determine where students are in their progress toward meeting the complex PEs. The *NGSA* design process can be used to unpack and identify the meaningful parts of the PEs that will be suitable for classroom-based assessment. The process emphasizes using the meaningful parts to construct comprehensive sets of smaller performance statements that we call *learning performances*. Learning performances are intermediary performance targets for instruction and assessment that can signal whether students are moving along a productive path to proficiency with a PE or PE bundle.

Learning Performances Help Instruction and Assessment to Work Together

Both instruction and assessment must work together to effectively help students build proficiency with the *NGSS* PEs over time (National Research Council, 2014). Instruction should help students build toward the PEs so that they can achieve the PEs for their grade level or grade band by the end of instruction. In parallel, assessment should provide a window into students' progress in building toward them. In the *NGSA* approach to assessment design, we use learning performances as three-dimensional learning goals that are smaller in scope than the performance expectations. These learning performances take the form of knowledge-in-use statements that incorporate aspects of DCIs, SEPs, and CCCs that students need to develop understanding of as they progress toward achieving a single PE or bundle. Learning performances are expressed using similar language as PEs and emphasize knowledge-in-use, but they are crafted at a more specified and manageable grain-size for classroom assessment purposes (Harris, Krajcik, Pellegrino & DeBarger, 2019). For example, the learning performance from which *Miranda's Mystery Liquids* (Figure 2.1) was developed is: *Students analyze and interpret data to determine whether substances are the same or different based upon patterns in characteristic properties.* What is foremost to call attention to is that this learning performance covers part of the multidimensional terrain that resides under the surface of the PE. Learning performances are useful in part because they are assessable in the midst of instruction, meaning that they provide opportunities to assess students as they are developing proficiency toward the larger PE.

Reexamination of the Example Task

Now that we have considered the role and utility of learning performances as classroom-based markers for monitoring students' progress toward meeting complex multidimensional PEs, consider *Miranda's Mystery Liquids* and those three questions again:

1. To what extent does the task match with the dimensional elements of the PE?

2. What type of information does this task stand to provide about students' proficiency with the PE?

3. How would this task be helpful in monitoring students' progress in building toward the PE?

Look carefully at the features of the task and how you responded to the task's prompts in light of the PE and also the learning performance. Although the task does not align fully with all parts of the PE, it does align with the parts of the PE that are represented in the learning performance. This learning performance describes an important three-dimensional requirement of the PE that students would need to achieve at some point during instruction as they progress toward meeting all that underlies the PE. Accordingly, the task stands to provide evidence on whether and how students are able to use their knowledge as specified in the learning performance. Such evidence is of high value because it enables us to assess "building toward" the PE.

Now that we have described our rationale for creating intermediary targets (i.e., learning performances) around which assessment tasks can be developed, we are ready to describe our process. In the remainder of this chapter, we provide an overview of the *NGSA* design process which can be used to construct instructionally supportive tasks that assess for three-dimensional learning. The process is introduced here and then explained in further detail in Chapters 3–7.

The *NGSA* Design Process: An Overview

Drawing conclusions about student performance from an assessment is essentially an effort of evidence-based argumentation. Assessment experts Robert Mislevy and colleagues (Mislevy, Steinberg & Almond, 2003; Mislevy & Haertel, 2006) used this idea as the basis for creating an assessment design framework called evidence-centered design (ECD). The ECD framework emphasizes the value of starting with a learning goal and determining the evidence that you would look for to make a judgment about students' performance of that learning goal, then specifying the features of tasks that will best bring out the evidence of performance. From the ECD perspective, the end goal of assessment is to make a claim about what students know and can do. In this endeavor, you must gather evidence to support that claim. This evidence typically takes the form of what students say, write, draw, or do in response to an assessment task.

Since it was proposed two decades ago, ECD has gained widespread attention in education as a worthwhile framework for principled assessment design. Noteworthy is that shortly after the release of the *NGSS*, the National Research Council released a comprehensive report on developing assessments for the *NGSS* (NRC, 2014). The report recommends that assessment design approaches follow the reasoning of ECD to ensure that tasks measure what matters for three-dimensional learning. The argumentative reasoning of ECD is integral to the *NGSA* design process for developing tasks that will provide evidence of students' three-dimensional performance as they build toward meeting the PEs.

The *NGSA* design process, illustrated in Figure 2.4, provides guidance on how to use PEs as the starting point for developing three-dimensional assessment tasks that can be used in the classroom to inform *NGSS* teaching and learning. This process allows us to create a set of learning performances from a PE or bundled group of PEs in a principled way. We then use the learning performances to guide the development of assessment tasks and accompanying rubrics. The process involves six major steps across three phases. The first phase, Steps 1–3, involves selecting a PE or PE bundle and systematically unpacking the dimensions to understand the assessable components. The elaborations from the unpacking are used to create a visual representation in the form of a map that lays out the dimensional "terrain" for fully achieving the PE or bundle. We refer to the map as an *integrated dimension map*. The second phase, Step 4, entails using the integrated dimension map in tandem with the unpacking to articulate and refine a set of learning performances that describe the proficiencies that students will need to demonstrate over time as they progress toward achieving the more comprehensive PE or PE bundle. The third phase, Steps 5–6, involves an organizational strategy called *design blueprints* for using learning performances to construct assessment tasks. Design blueprints provide the essential technical information for developing tasks that assess for three-dimensional learning. A single blueprint describes all of the major design decisions for creating one or more assessment tasks along with rubrics that each align to a learning performance.

FIGURE 2.4. The six steps of the *NGSA* task design process

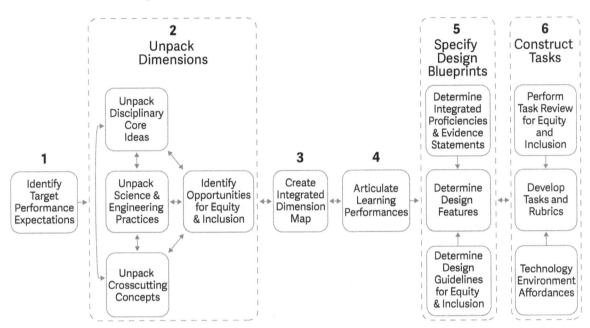

The Important Role of Equity and Inclusion in the *NGSA* Design Process

It is widely recognized today that designing and enacting three-dimensional assessments that promote equity and inclusion are a responsibility of all science educators. In *NGSS* classrooms, instruction and assessment should work together to attract all students to science learning and support and sustain their motivation and engagement. In the case of assessment, tasks should be designed and used in a way that both leverages and values students' background knowledge and experiences and connects them to rigorous science learning. When we accomplish this, we create the opportunity for the full range of students to demonstrate their three-dimensional learning, particularly those whose backgrounds have been underrepresented in science education and careers including girls, students who identify as LGBTQ, students with disabilities, underrepresented students of color, students with low-income family backgrounds, and emerging multilingual learners.

Equity and inclusion considerations are woven throughout the *NGSA* design process, beginning with the initial steps of selecting and unpacking the PEs and continuing through to the final products of tasks and rubrics. To support three-dimensional assessment design that promotes equity, guidance is provided for developing tasks that use

scenarios where the phenomena are likely to have broad interest or universal relevance for students, explicitly attend to language, and include scaffolds to make expectations explicit for students. To promote inclusion, guidance is provided for how to reduce bias in tasks, eliminate barriers that may interfere with student sensemaking, and support engagement so that students will be more likely to persist in reading and responding to tasks. When both equity and inclusion considerations are made, it becomes clear that assessment must allow for multiple ways for students to demonstrate their developing proficiencies and that using traditional or narrow assessment techniques and formats will not create an opportunity for students' multiple modes of performance to be recognized. The *NGSA* design process brings equity and inclusion considerations to the forefront so that tasks are accessible and fair for a wide range of students with varying backgrounds, skills, and abilities.

The Six Steps of the *NGSA* Design Process

What follows is a run-through of the six steps of the *NGSA* design process. Table 2.1 summarizes the key components of each step.

Step 1: Select Performance Expectation(s)

The first step in the design process is to select a target PE or a coherent bundle of PEs appropriate for classroom instruction. An appropriate PE or PE bundle should match with instruction such that the sequence of lessons and activities will provide an opportunity for students to build the knowledge and capabilities required by the PE or PE bundle over time. When there is a strong match between the PEs and the instruction, then it is suitable for the PE or PE bundle to serve as the focal point for developing three-dimensional assessment tasks.

Step 2: Unpack the NGSS Dimensions

One of the major pitfalls in designing assessment tasks for the *NGSS* is mistakenly believing that the one-sentence PE statement can be fully understood just by reading it. Similar to the iceberg (Figure 2.3), the PE in its entirety is very deep, and as designers we need to understand all the parts that are involved. Unpacking is a foundational step for detailing all that lies beneath the surface of a PE or PE bundle. The unpacking step is of high value because it enables designers to identify and elaborate on all that is involved for students to successfully demonstrate integrated proficiency. Importantly, unpacking focuses attention on the specific and meaningful aspects of the three dimensions, as well as the knowledge and capabilities that students need to develop along each of those dimensions. When information from a careful and thorough unpacking is documented, it becomes an important resource that can be used time and again during the design process to support and verify key design decisions.

Unpacking helps designers develop deep knowledge of the aspects of proficiency inherent within each dimension. It also focuses attention on identifying potential intersections between dimensions, thereby providing the foundation for integrating them later in the design process. Noteworthy too is that when unpacking the dimensions, it is of value to consider other SEPs and CCCs that could productively work together along with the DCI to build toward the PE or PE bundle. These considerations play an important role in the upcoming steps of mapping the dimensions and articulating learning performances.

When unpacking the dimensions, it is of paramount importance to always unpack with the student in mind. This includes describing students' prior knowledge and identifying likely student challenges with the dimensions; defining boundaries of what students should know and be able to do; earmarking issues of equity and inclusion that are relevant to the dimensions; identifying candidate phenomena relevant to both the PE and students' everyday lives and interests; and sketching out possible realistic scenarios that can provide a motivating context for making sense of phenomena.

As an example of unpacking the dimensions, consider PE MS-PS1-2 shown in Figure 2.2, which uses the SEP Analyzing and Interpreting Data, the elements from DCI PS1.A Structure and Properties of Matter, and the CCC Patterns. The DCI dimension is expressed within the portion of the PE focusing on the "properties of substances before and after the substances interact to determine if a chemical reaction has occurred." This part of the PE contains a considerable amount of disciplinary knowledge about the structure and properties of matter, thereby requiring that students must grasp a number of major ideas such as the following:

- Substances and their properties.
- How substances can interact.
- That new substances can be made from original substances.
- That changes in properties of substances from before they interact to afterward can serve as evidence that new substances are formed.

If we take a deeper dive and examine the two DCI elements underlying this PE (see the foundational boxes in Figure 2.2), several more ideas become visible, including:

- Pure substances have characteristic physical and chemical properties that can be observed and/or measured.
- Characteristic properties can be used to identify a substance and distinguish it from other substances.
- Substances react chemically in characteristic ways with other substances.
- During a chemical process, atoms in a substance can rearrange or regroup in new ways to form different molecules.

Important to note is that unpacking is much more involved than simply listing the various ideas found in the elements of the DCI. Unpacking also involves determining what students at the grade level are expected to know about the idea. In this case, we are considering middle school students. As an example, let's further unpack the last bullet, "During a chemical process, atoms in a substance can rearrange or regroup in new ways to form different molecules." To grasp this sub-idea, a middle school student would have to know that a molecule of a substance is always made up of the same type and number of atoms and that during a chemical reaction, the atoms can rearrange into new molecules to form new substances. These new molecules will always have the same type and number of atoms, just like the substances before the chemical reaction occurred. In this way, unpacking helps to establish the boundaries for the range and depth of DCI knowledge that students must attain as they build toward the PE. Also important is determining the prerequisite knowledge students should be expected to already possess, as well as any knowledge that goes beyond what students should know at the grade level. Closely related, it is important to identify known student challenges. In the case of DCI PS1.A, students typically have a difficult time visualizing how the arrangement and motion of atoms at the microscopic level can explain what is observed at the macroscopic level.

Another activity during the unpacking of a DCI is to identify phenomena that students can engage with that align with the DCI elements. Determining candidate phenomena and sketching possible scenarios to use in assessment tasks is an important part of unpacking that will directly inform task design. An engaging phenomenon or complex problem in an assessment task sets the stage for how students will engage with the task and provides a meaningful context for students to make sense of the task requirements. Importantly, a rich phenomenon and scenario can activate the appropriate knowledge and capabilities required for working through the task and demonstrating performance. To accomplish this, the phenomenon, or problem, needs to be relevant and engaging for students. Students should be able to use elements of the DCI with SEPs and CCCs to reason through and figure out the phenomenon or problem at hand, just as any scientist would. Ideally, the phenomenon or problem will allow students to engage not with "someone else's" science, but with science that they see as relevant to their own lived experiences. Further, these phenomena should be contextualized in compelling scenarios so that they create a "need-to-know" for students, and they should be complex enough that students must use all three dimensions to make sense of them. Achieving these aims, particularly while also ensuring the phenomenon or problem can be explained or solved using grade-appropriate knowledge and practices, is no small feat.

In tandem with the unpacking of the DCI, the SEP and CCC are also unpacked. When unpacking an SEP, the focus is on clearly articulating the essential grade band appropriate performance for the SEP. This includes specifying the aspects of the SEP that students are to perform as well as specifying the evidence required for students to demonstrate a high level of proficiency, identifying prior knowledge that is required of students to demonstrate the SEP, and identifying common challenges that students may encounter as they are developing sophistication with it. We also identify productive intersections between the SEP and other SEPs that are relevant for building toward the PE or PE bundle. When unpacking the CCCs, the focus is on identifying the important aspects of the CCC, as well as how the CCC intersects with both the SEP and particular sub-ideas of the DCI. Similar to unpacking SEPs, it is important to identify common challenges and to specify the evidence required for a student to demonstrate a high level of proficiency with the CCC.

The unpacking of all three dimensions is foundational for designing good assessment tasks and accompanying rubrics. Importantly, the unpacking step can promote consistency in the use of the dimensions throughout the design process. Detailed guidance for how to unpack is presented in Chapter 3.

Step 3: Map the Dimensions

The third step in the *NGSA* design process entails using the dimension elaborations from the unpacking to create what we call an *integrated dimension map* that provides a visual representation of the target PE or bundle. Mapping synthesizes the information from unpacking to lay out the DCI terrain and visually represent the most salient and productive intersections across the DCI, SEP, and CCC. The end result of mapping is an integrated dimension map that expresses the key relationships between the DCI elements that were elaborated in the unpacking and identifies how aspects of the SEP and CCC (also elaborated in the unpacking) can work with these disciplinary relationships to promote students' integrated proficiency.

TABLE 2.1. Key Components of the *NGSA* Design Process

Key Design Components

1. Performance Expectations

- Comprehensive knowledge-in-use learning goals to be achieved *by the end* of instruction
- Starting point for the *NGSA* Design Process
- An appropriate PE or PE bundle will match with instruction that aims to build the knowledge and capabilities required by the PE or PE bundle over time.

2. Unpacking

- Lays out the essential aspects of each dimension and identifies productive intersections within and across dimensions
- Promotes consistency in the use of dimensions during task design
- Documents equity and inclusion considerations such as students' prior knowledge, anticipated challenges, and language and literacy demands
- Identifies phenomena relevant to both the PE and students' interests as well as relatable scenarios that connect with students' everyday experiences

3. Integrated Dimension Maps

- Visual representation of the essential dimension elements and their sub-idea relationships
- Shows the area and boundaries of a performance expectation or bundle

4. Learning Performances

- Knowledge-in-use statements that take on the three-dimensional structure of a PE but are smaller in scope and align to only a portion of the PE
- Function in relation to other learning performances to cover the entire "terrain" of a PE or bundle
- Provide intermediary performance goals for creating assessment tasks for classroom use

5. Design Blueprints

- Guide the principled development of tasks aligned to learning performances
- Identify what should be included or not included in tasks and ensure that critical specifications are used consistently
- Specify the essential features that all tasks must include and the variable features that can vary among tasks
- Set down equity and inclusion guidelines so that tasks are accessible and fair for a wide range of students
- Include the integrated proficiencies and evidence statements that demonstrate performance

6. Three-Dimensional Tasks & Rubrics

- One or more tasks can be created from a single design blueprint
- Task should include a phenomenon that is contextualized in a scenario that encourages students to engage with the phenomenon and work through the task
- Responding to the task should require students to use the aspects of the SEP, DCI, and CCC targeted by the learning performance
- An integrated three-dimensional assessment task requires an integrated three-dimensional rubric

To develop an integrated dimension map, we start by laying out the major science sub-ideas of the DCI that was unpacked and specifying the relationships between them. Similar to a concept map, the mapping initially takes the form of a diagram that depicts the key sub-ideas with arrows that show the relationships among those sub-ideas. Figure 2.5 illustrates the mapping of the sub-ideas and grade-appropriate relationships for two bundled PEs, MS-PS1-2 and MS-PS1-5. These PEs are ideal to bundle together because they are complementary in their focus on chemical reactions and would likely be taught and assessed within the same instructional sequence or unit.

- MS-PS1-2. Analyze and interpret data on the properties of substances before and after the substances interact to determine if a chemical reaction has occurred.
- MS-PS1-5. Develop and use a model to describe how the total number of atoms does not change in a chemical reaction and thus mass is conserved.

FIGURE 2.5. A dimension map showing the elements of the disciplinary core idea(s), and how they relate to each other, for PEs MS-PS1-2 and MS-PS1-5

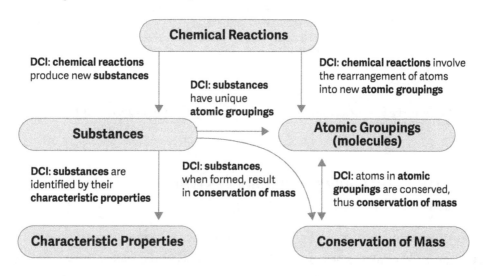

Once the key science sub-ideas of the DCI are laid out, we overlay the relevant SEPs and CCCs onto the map as shown in Figure 2.6. We add only the SEPs that students could use to demonstrate their understanding of the DCIs and CCCs. For instance, the SEP Computational Thinking, while an important practice, does not work as well with these sub-ideas as it might with others. Next, we add CCCs that students could engage with and connect to the DCIs, or that might arise from their engagement with the SEP. We explicitly represent and consider each of the three dimensions and how they relate to one another in this step.

This theme of integration occurs throughout the design process to ensure that all three dimensions are incorporated into the tasks and rubrics.

FIGURE 2.6. An integrated dimension map showing the relationships between the elements of the disciplinary core idea(s), crosscutting concepts, and science and engineering practices for the bundled PEs MS-PS1-2 and MS-PS1-5

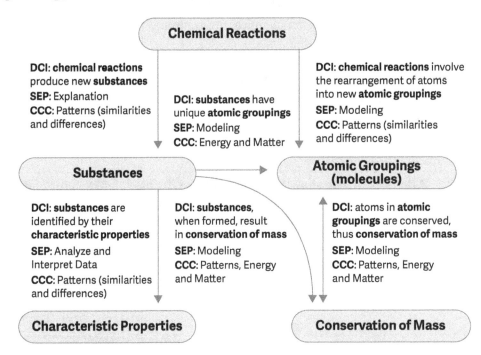

Step 4: Articulate Learning Performances With Multiple Opportunities for Access

The fourth step involves using the integrated dimension map as the starting point for articulating and refining a set of learning performances that collectively describe the proficiencies that students need to demonstrate in order to meet a PE or PE bundle. A single learning performance is crafted as a knowledge-in-use statement that is smaller in scope and covers a designated area of an integrated dimension map. Like a PE, each learning performance has a three-dimensional structure that helps maintain the requirement that students use their knowledge to make sense of phenomena or solve problems.

The integrated dimension map shown in Figure 2.6 was used to create the learning performance we introduced earlier in the chapter. This learning performance was created from the lower left region of the map where the major ideas of *substances* and *characteristic properties* are linked. The two major ideas are linked by a DCI sub-idea, the SEP Analyzing and Interpreting Data, and an aspect of the CCC Patterns. The components from this area of the map were brought together to develop a three-dimensional learning performance. We started with the action of the SEP and then added the DCI elements and the CCC to arrive at the learning performance: Students analyze and interpret data to determine whether substances are the same or different based upon patterns in characteristic properties.

You likely noticed that the learning performance we specified covers a relatively modest portion of the integrated dimension map. This same map can be used to articulate multiple learning performances for the PE bundle. All told, we articulated six learning performances for the bundled MS-PS1-2 and MS-PS1-5, one of which is the learning performance we have been exploring. As a rule, the set of learning performances should collectively describe the proficiencies that students need to demonstrate in order to meet a PE bundle. In this way, learning performances are intended to function together *as a set* that covers all areas of an integrated dimension map. An individual learning performance covers a smaller area; all the learning performances together address the entire PE bundle. Chapter 4 provides pointers on how to create sets of learning performances.

Another noteworthy feature of a learning performance is that it can use different SEPs and CCCs in concert with elements of the DCI. For instance, the target performance expectation might use one SEP (e.g., Analyzing and Interpreting Data) and one CCC (e.g., Patterns) but the learning performances could use alternate SEPs (e.g., Constructing Explanations) and CCCs (e.g., Cause and Effect). This integration of other complementary SEPs and CCCs is aimed at providing students with varied opportunities to engage with elements of the DCI in meaningful ways. A benefit of this feature is that it allows for multiple modes of performance by students to demonstrate their proficiencies. Also, there can be a real learning benefit for varying the SEPs. For example, some SEPs have greater linguistic demands (e.g., Constructing Explanations, Engaging in Argument from Evidence) and stand to provide students, including those with emerging language and comprehension skills, with increased opportunities to practice and develop their linguistic capacities, thereby promoting all students' learning (NGSS Lead States, 2013, Appendix D). Further, integrating DCI elements with various SEPs and CCCs supports students in flexible use of all three of the dimensions, providing for increased opportunities to develop their capabilities for using and applying knowledge to make sense of phenomena or solve problems.

A single learning performance is a claim that we make about what students know and can do regarding an essential part of a PE or PE bundle. Because learning performances specify what students should know and do, they serve as the keystone of our assessment argument. We use a learning performance to design one or more assessment tasks, which provide observable evidence (in the form of student responses) that we can use to judge whether the student can use the knowledge in the learning performance. For example, students' responses to the task *Miranda's Mystery Liquids* can be used as evidence of proficiency in using and applying the knowledge of the learning performance.

Step 5: Specify Design Blueprints

In this step, we use an organizational strategy called *design blueprints* to guide the principled development of tasks. A design blueprint is a document that brings together all that must be taken into account when creating tasks for each learning performance. The value of using blueprints is that they can make implicit design decisions *explicit* for those who will be doing the actual constructing of tasks and rubrics. Importantly, blueprints set the boundaries for what should be included or not included in tasks and they ensure that critical specifications, like task features, scaffolding levels, and format types are used consistently. When designers construct one or more assessment tasks for the same learning performance, using design blueprints increases the likelihood that the tasks will include all of the essential features from the unpacking and mapping, and that they will elicit the intended evidence from students. On the other hand, when task designers do not follow a set of specifications, important features may be overlooked and the end result may be tasks that do not measure what was intended. Also, when tasks are constructed primarily based on the designer's intuitions about what makes "good" tasks, that knowledge resides with the individual and may not always be readily recalled or consistently applied. Moreover, the designer's intuition cannot be efficiently taught nor reliably replicated by others who are not aware of the tacit knowledge held by the assessment designer.

When specifying a blueprint, we concern ourselves primarily with documenting and making explicit our thinking about what observable evidence the task will provide, why and how the task will provide that evidence, and why that evidence is relevant to the claim (i.e., the learning performance). To document this thinking, we create a design blueprint that is used to specify features needed in a task so that it aligns with the learning performance. In the end, each learning performance will have a design blueprint that outlines various features of a task that a designer needs to consider when developing any task that aligns to the learning performance. Think of a design blueprint as a guide with directions to follow in developing tasks. A value of design blueprints is that they can be used to make not just one task, but multiple tasks that are each aligned to the same learning performance.

When producing blueprints, we organize the specifications around the following five practical design questions:

1. What should students know and be able to do in order to demonstrate this learning performance?

2. What could students say or do that would provide evidence that they can use the knowledge described in this learning performance?

3. What are the essential features that all tasks constructed for this learning performance will need to include?

4. What are the variable features that some tasks (but not all tasks) constructed for this learning performance will need to include?

5. What equity and inclusion considerations are necessary to include to ensure that tasks constructed for this learning performance will be accessible and fair for a wide range of students?

As we answer these questions, we document our responses and label them as *integrated proficiencies, evidence statements, essential task features, variable task features,* and *equity and inclusion considerations* respectively. *Integrated proficiencies* refer to the abilities that students must employ to demonstrate the learning performance. For each integrated proficiency, we identify the observable evidence that students need to provide in order to show that they can use and apply the knowledge in the learning performance. We refer to these as *evidence statements* and they are the second key component of our "assessment argument." Importantly, they provide direct guidance for what we should look for in a student's performance. Because learning performances are multidimensional, our evidence statements are also multidimensional. An evidence statement succinctly describes the information we need to observe in a student's response to a task so that we can defend the claim made about what that student knows and can do.

An assessment should prompt students to say, do, or create something with the aim of getting them to provide observable data evidence that demonstrates their current level of integrated knowledge use of the DCIs, CCCs, and SEPs. For example, for the learning performance, *Students analyze and interpret data to determine whether substances are the same or different based upon patterns in characteristic properties,* the integrated proficiencies might include:

A. Ability to analyze data on substances to identify patterns in characteristic properties.

B. Ability to describe that two or more unknown substance samples are the same substance or different substances based on patterns in characteristic properties.

The corresponding evidence statements might be:

A. Students identify patterns in data regarding the characteristic properties (e.g., density, melting point, boiling point) of substances.

B. Students use evidence from data to write a conclusion with reasoning that if the same properties are found in multiple unknown samples, then those samples are the same substance because they share all of the same characteristic properties.

You can learn more about integrated proficiencies and evidence statements, as well as how to write them, in Chapter 4.

Next, we describe essential and variable task features. Think of *essential task features* as the features that all tasks need to include to meet a particular learning performance. For instance, all tasks constructed for the learning performance we are discussing must have a scenario that presents samples of substances that can be identified using characteristic properties at the middle school level. On the other hand, *variable task features* specify the important features that can vary in some form or by option across tasks. For meeting our learning performance, the type of substance samples and their states—whether gases, liquids, or solids—could vary across different tasks and yet still meet the requirements of the evidence statements. A helpful analogy that can further distinguish essential and variable task features is car manufacturing. When manufacturing cars, all cars that come off the assembly line must have an engine that will provide the energy for motion. Accordingly, the engine is an essential feature that all cars must share. However, the type of engine (e.g., fuel engine or electric motor) can differ and we still have a car. So, the engine type is a variable feature among cars. Describing the essential and variable task features provides the third key component of a coherent assessment argument. This component enables the development of multiple tasks that all share the same essential features yet may still differ in ways that matter for when and what purposes you might use them. Chapter 5 provides further details on identifying and using essential and variable task features.

Design blueprints also include *equity and inclusion considerations* that make explicit how tasks will leverage and value students' background knowledge, capabilities, and experiences. Our emphasis is on developing tasks that enable all students to show how they can use and apply what they know. To achieve this, designers must give attention to a range of matters, including considering language and sentence structure, including ways to reduce bias and increase inclusiveness, identifying phenomena that are relevant to students' lives, and incorporating features that enable a mixture of modes for demonstrating performance. Chapter 5 provides guidance on determining equity and inclusion guidelines and incorporating them into design blueprints.

Step 6: Construct Tasks and Rubrics

Once all the specifications (i.e., integrated proficiencies, evidence statements, essential task features, variable task features, and equity and inclusion considerations) are organized into a design blueprint, the final step is to construct the assessment tasks and accompanying rubrics (see Chapters 5 and 7). Writing a task involves selecting a compelling phenomenon or problem, describing or representing that phenomenon or problem in a scenario, writing the task prompts, and defining how students with different levels of proficiency will respond to the task (part of which involves constructing a rubric). It is imperative to select a phenomenon or problem that will allow students to make use of their knowledge and demonstrate the extent to which they can use their knowledge as required by the learning performance. A well-written scenario will introduce the phenomenon and help students home in on what they need to figure out or explain. Together, the scenario and phenomenon should create a compelling need for students to engage in reasoning about that phenomenon and work through the task.

The design blueprint serves as a guide in writing tasks. When constructing a single task, referring back to the design blueprint is critical to ensure the task can provide the evidence needed to make an argument that the learner has met the knowledge encompassed in the learning performance. When designing multiple tasks from a blueprint, each task should fully stand alone in representing the learning performance.

The task prompts should be written in a manner that elicits an integrated three-dimensional response, and responding to the task should require students to use the target aspects of the SEP, DCI, and CCC of the learning performance. For instance, *Miranda's Mystery Liquids* (see Figure 2.1) is a task that requires students to use all three dimensions in an integrated manner. In this task, students are first asked to consider what data they could use to identify which samples are from the same substance. To be able to do so, students need to know that density and boiling point are characteristic properties, that mass and volume are not characteristic properties, and that characteristic properties are unique to a substance. These ideas are related to the DCI and necessary for applying the SEP. The second prompt asks students to make a conclusion about which substances are the same. To successfully respond, students need to analyze the data in the table (the SEP), look for relevant patterns (the CCC), and identify which samples could be the same substance because of their properties (the DCI). They must use the dimensions in an integrated manner to make a conclusion that liquids in samples 2 and 4 are the same because they have similar densities and boiling points. As such, students make use of all three dimensions of scientific knowledge, providing evidence that they can use the knowledge in the ways specified in the learning performance.

Once the task is written in a manner that requires students to use all three dimensions, we develop a rubric that can be used to evaluate student responses (see Chapter 7). In short, an integrated three-dimensional assessment task requires an integrated three-dimensional rubric. The rubric needs to provide information on all three dimensions. A good rubric provides the teacher and student with useful, multidimensional information about a student's proficiency. To develop rubrics, we need to make use of the integrated proficiencies and evidence statements identified as part of writing the task design blueprints (Step 5). The rubrics are constructed using multiple parts, with each part representing a different integrated proficiency. Crucially, these rubrics preserve the multidimensional nature of the integrated proficiencies, meaning that the three *NGSS* dimensions are not separated into different rubrics.

Another important activity in this design step is to conduct a careful review of the task before it is finalized and deemed ready for classroom use. This review includes making sure that the task meets all of the design specifications in the blueprint and that it is accessible and fair to a broad range of students. Regarding the latter, we conduct an equity and inclusion review to ensure that the assessment task values students' background knowledge and experiences, and that the task enables all students to show how they can use and apply what they know. Following the review, we modify tasks as needed to address any problematic issues that were flagged.

When using assessments that are intended to be instructionally supportive, it is generally not useful to only determine whether students can or cannot do something. It is most useful to know what their performance looks like, the varying levels of performance among students, and how you might build on these performances over time (see Chapters 7 and 8). To this end, the *NGSA* design process can be used to develop task rubrics that determine the range of students' three-dimensional performances within a given classroom and provide guidance for next instructional steps (see Chapter 7).

Primary Takeaways

In this chapter, we introduced a process for systematically designing assessment tasks that can be used in the classroom to inform *NGSS* teaching and learning. The assessment design process is of high value because it enables various designers—whether classroom teachers, state and district educational leaders, or curriculum and assessment developers—to develop three-dimensional tasks and accompanying rubrics that are instructionally supportive. The primary takeaways from this chapter are:

1. Both instruction and assessment must work together to effectively help students build proficiency with the *NGSS* PEs over time. Instruction should help students build toward the PEs, and assessment tasks for *NGSS* classrooms need to be three-dimensional so that they can provide information about how students are progressing toward achieving them.

2. The *Next Generation Science Assessment* (*NGSA*) design process provides step-by-step guidance on how to use PEs as the starting point for developing three-dimensional assessment tasks. The process involves six major steps with equity and inclusion considerations woven throughout. It brings equity and inclusion considerations to the forefront so that tasks are accessible and fair for a wide range of students with varying backgrounds, skills, and abilities.

3. The *NGSA* design process can be used to unpack and identify the meaningful parts of PEs that will be suitable for classroom-based assessment. The process emphasizes using the meaningful parts to construct comprehensive sets of smaller performance statements that are called *learning performances*. Learning performances are intermediary performance targets for instruction and assessment that can signal whether students are moving along a productive path to proficiency with a PE or PE bundle.

4. Design blueprints are an organizational strategy for the principled development of tasks aligned to learning performances. Clearly specified design blueprints can bring coherence and consistency to the work of creating three-dimensional tasks. They serve as the ground plans for task designers, providing clear information for creating tasks and common reference points for checking work.

5. An integrated three-dimensional assessment task requires a corresponding integrated three-dimensional rubric. A well-designed rubric can illustrate what

three-dimensional science performance looks like and provide guidance to inform subsequent instruction that will help move students forward in building toward the PEs.

In Chapters 3–7, we delve deeper into the *NGSA* design process and provide step-by-step guidance for constructing well-aligned tasks that can be used in the classroom to follow and support students' progress in three-dimensional learning.

References

Harris, C. J. et al. 2019. Designing knowledge-in-use assessments to promote deeper learning. *Educational Measurement: Issues and Practice,* 38(2), 53–67.

Mislevy, R. J. & G. D. Haertel. 2006. Implications of evidence-centered design for educational testing. *Educational Measurement: Issues and Practice*, 25(4), 6–20.

Mislevy, R.J., L. S. Steinberg & R. G. Almond. 2003. On the structure of educational assessments. *Measurement: Interdisciplinary Research and Perspectives*, 1(1), 3–62.

National Research Council (NRC). 2012. *A framework for K–12 science education: Practices, crosscutting concepts, and core ideas*. Washington, DC: National Academies Press.

National Research Council (NRC). 2014. *Developing Assessments for the Next Generation Science Standards*. Washington, DC: National Academies Press.

NGSS Lead States. 2013. *Next Generation Science Standards: For states, by states*. Washington, DC: National Academies Press.

CHAPTER 3

Unpacking and Mapping the *NGSS* Dimensions

Sania Zahra Zaidi, University of Illinois Chicago • Kevin W. McElhaney,
Digital Promise • Nonye Alozie, SRI International • Joseph Krajcik,
Michigan State University

The *NGSS* Performance Expectations: Just the Tip of the Iceberg

Have you ever tried to use the *NGSS* performance expectations (PEs) to guide instruction or assessment in your classroom? If so, you have probably realized that the PEs are far more complex than meets the eye. It can be helpful to think of a concise PE statement as just the tip of the *NGSS* "iceberg." Under the surface, each PE encompasses a vast domain and a wide range of proficiencies, which are impossible to address in a single lesson or assess using a single assessment task. This inherent complexity can make it challenging to build students' three-dimensional proficiency with the PEs through instruction and assessment, and raises an important question: *How can you and other teachers deconstruct and make sense of performance expectations to accurately determine what students should know and be able to do?*

Answering this question requires a design approach that can systematically deconstruct and identify the meaningful parts of the PEs, while also bringing these parts back together in a way that preserves their three-dimensionality. In this chapter, we describe two foundational steps of the *NGSA* design process: (1) unpacking, through which we deconstruct the three dimensions of the *NGSS* PEs, and (2) mapping the dimensions, through which we synthesize the information from the unpacking to create a visual representation.

Unpacking allows us to identify the knowledge and capabilities students need to develop along each of the three dimensions to attain integrated proficiency with a PE or PE bundle. Through unpacking, we can understand the key aspects of each of the three dimensions at a particular grade level or grade band. Mapping then allows us to bring the dimensions back together by identifying how the dimensions are interrelated and intended to work together to promote students' integrated proficiency.

In this chapter, we use examples from several middle school PEs to illustrate unpacking and mapping, highlighting important aspects that emerge in each step. Unpacking and mapping the dimensions provide the underpinning for crafting learning performance statements (see Chapter 4) and creating assessment tasks (see Chapter 5). We close the chapter by summarizing how unpacking and mapping are essential in creating instructionally supportive assessment tasks that elicit the evidence of students' three-dimensional proficiency that we seek as educators.

Why Unpack and Map the *NGSS* Dimensions?

The unpacking step reveals the complexity that lies below the surface of the PE. The text of a one-sentence PE statement does not make clear all the underlying knowledge and capabilities required for students to demonstrate integrated proficiency, nor does it immediately reveal the boundaries of what students should know and be able to do at a particular grade. Unpacking helps designers to develop deep knowledge of the aspects of proficiency inherent within each dimension. When the information from a careful and thorough unpacking is documented, it becomes an indispensable resource that can be used time and again during the design process to support and verify key design decisions. For these reasons, the unpacking of all three dimensions is considered foundational for designing good assessment tasks and accompanying rubrics.

In designing tasks for assessing three-dimensional learning, we must also understand how the three dimensions of a PE relate to each other, since PEs are not arbitrary combinations of the dimensions. In fact, PEs are intended to integrate the dimensions in specific and meaningful ways. The mapping step directs attention to the important relationships among the three dimensions (and other related dimensions) by illustrating how science and engineering practices (SEPs) and crosscutting concepts (CCCs) integrate with the specific disciplinary relationships inherent in a bundle of PEs. Each map is a visual representation that lays out the multidimensional "terrain" for fully achieving the PE or PE bundle. Beneficially, mapping provides a vantage point for designers to see the range of possible intermediary performances—what we call learning performances—that students will need to demonstrate to show that they are progressing toward achieving the full breadth of the PE or bundle. Once designers reach consensus on the contents of a map, it is used in Step 4 of the *NGSA* process to articulate learning performances as three-dimensional performance statements that are smaller in scope than the PEs, for which specific assessment tasks can be designed for classroom use (see Chapter 4).

Getting Ready to Unpack and Map a PE Bundle

We organize for unpacking by identifying a coherent bundle of PEs appropriate for classroom instruction. An appropriate PE bundle will match with instruction such that the sequence of lessons and activities will provide an opportunity for students to build the knowledge and capabilities required by the PE bundle over time. Once the PEs have been selected, we get our footing with a careful examination of the disciplinary core ideas (DCIs), which are comprised of statements associated with the target PEs (referred to as "elements" in the *NGSS*); followed by the science and engineering practices (SEPs); and finally, the crosscutting concepts (CCCs) that are part of (or related to) the target PEs. Unpacking entails specifying the key aspects of each dimension for the target grade level, identifying the boundaries of what students must know and do (i.e., areas that are not the focus of learning for the target grade), and documenting the prerequisite knowledge and capabilities students must have to develop grade-level-appropriate understanding of each dimension. When unpacking SEPs and CCCs, we also identify related SEPs and CCCs that are not part of the target PEs but could still provide evidence of students' building proficiency with those PEs. Equity and inclusion considerations that are relevant to the dimensions are also documented. This includes, for example, attending to the diverse linguistic needs of students by considering the many different ways that students can engage with and use language when working with the three dimensions of the PE.

The act of unpacking is a "deep dive" that can generate a large amount of information, which in turn can quickly become unwieldy. For this reason, it is a good idea to plan ahead for how you will organize and present the information. We have found that the most efficient way to organize and present the comprehensive information from an unpacking is in a table format. The value of using a table format is that it can provide a structure for proceeding through the unpacking exercise. For example, a table with descriptive headings and strategically ordered guiding questions can provide a route to follow that will ensure that all of the critical key areas of unpacking are accomplished. Once the exercise is complete, the tables then serve as an accessible and comprehensive resource that can be used by you or any other designers onward through the *NGSA* design process. When

creating tables, we recommend that abbreviations and terminology should be used consistently; descriptive headings should be clear and brief; and guiding questions should strongly align with the headings. Worth mentioning is that space should be allotted for citing sources so that if needed at any time you can refer back to the original source material—whether, for example, the source is the *NGSS* Appendices or an NSTA publication such as *The NSTA Atlas of the Three Dimensions* (Willard, 2020).

After the unpacking exercise is completed, we then proceed to mapping, where we pull directly from the information compiled during unpacking to create a visual representation of the PE bundle that we refer to as an *integrated dimension map*. In this step, we use pencil and paper or visual design software, such as a drawing or presentation tool, to draw a diagram that expresses the key relationships between the DCI elements and the aspects of the SEPs and CCCs within a bundle. To develop an integrated dimension map, we start by laying out and connecting the sub-ideas of the DCI elements that were unpacked and the relationships between them. Similar to a concept map, the mapping initially takes the form of a diagram that depicts the key sub-ideas with arrows that show the relationships among those sub-ideas. Once the sub-ideas of the DCI are laid out, we integrate the SEPs and CCCs into the map to illustrate the various ways that the relationships between the DCI elements can work with aspects of the SEPs and CCCs to build students' integrated proficiency. We have found that it can take multiple iterations to develop a map that fully represents a given bundle and clearly illustrates the relationships and productive intersections. A helpful strategy is to use presentation software with features such as text boxes, shapes, and arrows that you can easily arrange. If designing collaboratively, presentation software will also support sharing and allow for annotated commenting. In this next part of the chapter, we provide details and guidance on how to bundle PEs and conduct the unpacking and mapping of a bundle.

Guidance for Bundling PEs

The *NGSA* design process always begins with selecting and bundling PEs that matter for instruction. A bundle consists of two or more PEs that are complementary, and when grouped together, they can be leveraged to make sense of phenomena. Importantly, a promising bundle of PEs for assessment design will also be found to work together within an instructional sequence or unit. Bundling with instruction in mind is recommended because many recent *NGSS*-designed curriculum materials follow the *NGSS* guideline to bundle multiple PEs in instruction, rather than to isolate PEs and teach them separately from one another.

While there is no one right way to bundle PEs, we have learned that PEs that share DCI elements can be good candidates for bundling. As a general rule, PEs can potentially be bundled when there is a clear connection between the candidate PEs and when they are emphasized in instruction together. The benefit of bundling related PEs and simultaneously unpacking them is that you can capitalize on the connections between PEs to coherently identify their unique and shared assessable components. For example, two related PEs that might be bundled for middle school physical sciences are within the topic of Matter and its Interactions (MS-PS1):

- MS-PS1-2: Analyze and interpret data on the properties of substances before and after the substances interact to determine if a chemical reaction has occurred.
- MS-PS1-5: Develop and use a model to describe how the total number of atoms does not change in a chemical reaction and thus mass is conserved.

These PEs are an ideal bundle because they are complementary in their focus on chemical reactions and would likely be taught and assessed within the same instructional sequence or unit. Both PEs are associated with the DCI of Matter and its Interactions and together highlight two components of the DCI: (1) PS1.A: Structures and Properties of Matter and (2) PS1.B: Chemical Reactions. In the *NGSS*, each DCI component is further broken down into elements, which can be associated with one or more PEs. For example, MS-PS1-2 includes an element within the DCI component PS1.A: Structures and Properties of Matter. In this case, the element indicates that each pure substance can be identified using its characteristic physical and chemical properties (for any bulk quantity under given conditions). Differently, MS-PS1-5 includes an element from the DCI component PS1.B: Chemical Reactions. This element states that the total number of each type of atoms is conserved in a chemical reaction, and thus the mass does not change. However, both PEs share another element from the DCI component PS1.B: Chemical Reactions. This element focuses on how substances react chemically to form new substances and, more specifically, how in a chemical process the atoms that make up the original substances are regrouped into different molecules, and these new substances have different properties than those of the reactants. Notably, both PEs include one unique DCI element related to matter and its interactions (i.e., one unique to each individual PE) and one shared between the two PEs. This connection between the PEs makes them advantageous for bundling.

Guidance for Unpacking the NGSS Dimensions

To understand what lies beneath the surface of a given PE, we must move beyond merely parsing the three dimensions in the PE and instead move toward understanding the multidimensional proficiency that students must develop over time. This includes the knowl-

edge and capabilities required to demonstrate proficiency. We must come to know what each dimension entails *and* how the dimensions are connected. Without a clear sense of how the dimensions are related, it will be difficult to envision how students can integrate the dimensions in meaningful ways to make sense of phenomena. Importantly for assessment design, we need to have this vision so that we can create tasks that will enable students to demonstrate their three-dimensional learning.

Unpacking Disciplinary Core Ideas

Unpacking begins with attention to the DCI components and elements that are included within your PE bundle. Identifying and elaborating the elements and their sub-ideas is typically the first undertaking, but it is important to emphasize that unpacking is much more than listing the various sub-ideas found in the elements of the DCI components. In the *NGSA* design process, the unpacking is conducted across five focus areas: (1) elaborating the DCI elements and sub-ideas, (2) defining boundaries of what students should know, (3) describing prerequisite knowledge, (4) identifying relevant phenomena to promote equitable access and inclusion for all students, and (5) identifying student strengths and challenges. Table 3.1 presents the five focus areas along with guiding questions for unpacking in each area.

TABLE 3.1. Focus Areas and Guiding Questions for DCI Unpacking

DCI Focus Area	Guiding Questions
1. **Elaborating the DCI Elements and Sub-ideas**	• What are the sub-ideas that are key to the element(s) of the DCI? • How are the sub-ideas related to each other and to the PE or bundle?
2. **Defining Boundaries**	• What areas of the DCI's element(s) lie outside the scope of learning at this grade band?
3. **Describing Prerequisite Knowledge**	• What knowledge and capabilities (both from this topic and from other topics) do students need to achieve an understanding of the element(s) of the DCI?
4. **Identifying Relevant Phenomena**	• What phenomena provide widely accessible examples or applications of the DCI elements and sub-ideas? • What relatable real-world situations could contextualize the phenomena and create a need for making sense of it?
5. **Identifying Student Strengths and Challenges**	• What everyday experiences have students had that would be relevant for this DCI? • What commonly held student ideas differ in important ways from the scientifically accepted understanding of the DCI?

DCI Focus Area 1: Elaborating the DCI Elements and Sub-ideas

When elaborating the DCI, we begin by examining the sub-ideas that are integral to the DCI elements associated with the PE bundle. The elements associated with PEs can be found as a bulleted list in the *NGSS* under the DCI foundation box pertaining to the target PE. Once we have identified the elements associated with the target PE, we determine the ways in which students will engage with the elements at a particular grade band. We then elaborate on the various sub-ideas contained within the DCI elements to understand how they relate to each other and to the PE. By describing the sub-ideas in detail, we ensure that we fully understand the scope of the DCI as it pertains to the PE. Additionally, the elaboration process takes into account grade level expectations, which ensures that we elaborate within the appropriate grade band. Helpful resources for elaborating on DCIs include the *Framework for K–12 Science Education* (NRC, 2012), the *NGSS* (NGSS Lead States, 2013), Duncan, Krajcik, and Rivet's (2017) book, *Disciplinary Core Ideas: Reshaping Teaching and Learning*, and the *NSTA Quick Reference Guide to the NGSS, K–12* (Willard, 2015).

Consider the example of elaborating on the elements of the DCI component PS1.B: Chemical Reactions (see Table 3.2). When elaborating the first element, we determine that it addresses a macroscopic view of chemical reactions. The second element, however, emphasizes the particulate view of chemical reactions. Thus, by elaborating on the meaning of the sub-ideas, we are able to capture the grade level expectation (e.g., understanding chemical reactions at both macroscopic and particulate levels) in the target PEs.

TABLE 3.2. Illustrative Example of Elaboration of the Elements of the DCI Component: PS1.B: Chemical Reactions Associated With PEs MS-PS1-2 and MS-PS1-5

Elements	Elaboration of the Elements
Substance reacts chemically in characteristic ways. In a chemical process, the atoms that make up the original substances are regrouped into different molecules, and those new substances have different properties from those of the reactants.	**Macroscopic level** A chemical reaction is the change of substance(s) into a new one that has a different chemical identity. It is usually observed through physical effects, such as the formation of a precipitate or gas, a color change, or heat transfer. However, the confirmation of chemical change can only be validated by analysis of the properties of the products. 1. When substances react chemically one or more new substances are formed. 2. New substances that form are the products of a chemical reaction. 3. It is possible for a single substance to undergo a chemical reaction. 4. Liquids, solids, or gases can be reactants or products in chemical reactions.

(Continued)

Table 3.2. *(continued)*

Elements	Elaboration of the Elements
The total number of each type of atom is conserved and thus the mass does not change.	**Particulate Level** A chemical reaction is a process in which the atoms that make up the molecules of the original substances rearrange into new molecules, so that the types and number of atoms do not change. 1. New substances are made of the same kinds of atoms as the original substances 2. The number of atoms before and after the substances interact are equal. 3. The total mass of the substances before and after the substances interact is equal, or the mass is conserved.

DCI Focus Area 2: Defining Boundaries for What Students Should Know at a Grade Band

When defining boundaries, we ask, "What areas of the DCI's element(s) lie within and outside the scope of learning at this grade band?" This question helps us to consider which sub-ideas within the DCI elements have aspects that were the focus of a previous grade band (lower boundary) or will be the focus of an upcoming grade band (upper boundary). By taking into account the lower and upper boundaries, we can better clarify what students are or are not expected to know. For example, when unpacking the DCI elements for chemical reactions (see Table 3.2) and defining their boundaries, it becomes clear that students are not expected to know *why* different atomic groupings occur and result in different properties, but only that substances with different atomic groupings have different properties and are therefore different substances. Information such as this is important to document to ensure that assessment tasks will not overreach or fall short in eliciting the knowledge that will be required of students. To the point, this focus area is meant to keep the elaboration of ideas at a grade-appropriate level for students.

As a resource, we recommend using the assessment boundaries listed for PEs in the *NGSS* to identify grade band expectations. Additionally, the DCI expectations for a grade band both below and above the target grade band found in the *Framework* (NRC, 2012) can help in identifying the lower and upper boundary for unpacking the target DCI for a given grade. The *NGSS*'s Appendix E: DCI Progression can also be a helpful resource for both elaborating on the ideas at a particular grade level and identifying the upper and lower boundaries of the elaborated ideas. Finally, another valuable resource is Willard's book, *The NSTA Atlas of the Three Dimensions* (2020).

DCI Focus Area 3: Describing Prerequisite Knowledge

After identifying, elaborating, and defining the boundaries of the DCI elements associated with the PEs, we address the question, "What knowledge and capabilities do students need to achieve an understanding of the element(s) of the DCI?" Prerequisite knowledge is comprised of (a) prior knowledge from previous grades, and (b) concurrently developing knowledge from the current grade, both of which are needed for students to attain proficiency with the DCI elements.

When documenting prerequisite knowledge, it is important to identify what you can expect students to already know from prior instructional experiences, as well as the knowledge students will be expected to develop at their current grade level for understanding the DCI elements. The benefit of this focus area is that it provides designers with a clear grasp of what knowledge and capabilities students will be expected to bring to an assessment task. With this information in hand, tasks can be thoughtfully designed to leverage that prerequisite knowledge. For example, regarding prior knowledge related to the DCI elements for chemical reactions, students should know by middle school that all matter is made of particles that are too small to be seen. In middle school, students will build upon this prior knowledge to concurrently develop knowledge about atoms and molecules and how their interactions at the microscopic level can explain chemical reactions that are observed at the macroscopic level. The implication for task design is that we can expect that students will have some prior knowledge about the particle nature of matter that they can apply to a given task. Information such as this can help designers determine the appropriate entry level for a given task and also inform the extent to which scaffolds will be used.

Documenting concurrently developing knowledge can also help designers to see how and where an assessment task could potentially fit within a sequence of instruction. For example, when identifying the concurrently developing knowledge for our bundled PEs MS-PS1-2 and MS-PS1-5, we find that DCI elements from MS-PS1-1: *Develop models to describe the atomic composition of simple molecules and extended structures*, are prerequisites for the PE bundle. Notably, students must first come to know that substances are made of atoms that form simple molecules or extended structures as a stepping stone for building their three-dimensional proficiency with the PE bundle that calls for using models to describe the conservation of atoms in a chemical reaction.

Resources to assist in documenting prior knowledge and concurrently developing knowledge include the grade level expectations for DCIs that can be found in the *Framework* (NRC, 2012) as well as the matrices describing DCI progressions across grades K–12 that are presented in Appendix E of the *NGSS* (NGSS Lead States, 2013). Another valuable resource is the *NSTA Atlas of the Three Dimensions* (Willard, 2020), which maps out learning progressions based on the *Framework* and the *NGSS*. *NGSS*-designed curriculum materials can also provide insight into prerequisite knowledge and the progression for how knowledge can be expected to develop over time.

DCI Focus Area 4: Identifying Relevant Phenomena

Another focus area in the unpacking of a DCI is to identify compelling phenomena that students can engage with that align with the DCI elements. Because explaining phenomena is a central goal of three-dimensional learning in *NGSS* classroom, assessments should be phenomenon-centered too. We have found that it is of value to generate a listing of candidate phenomena relatively early in the design process that you can return to later when developing tasks. When generating your list, we recommend that you consider phenomena that will provide widely accessible examples or applications of the DCI elements and sub-ideas. Alongside, it is beneficial to sketch out any relatable real-world situations that could contextualize a candidate phenomenon and create a need for making sense of it. Important, too, is to consider students' perspectives and experiences when choosing phenomena and sketching scenarios. Note that there can potentially be multiple phenomena that match with the DCI elements, but the aim is to identify those that can be explained or solved using grade-appropriate knowledge and practices *and* will be relevant and engaging for students. Ideally, the phenomena or problem situations will allow students to engage not with "someone else's" science, but with science that they see as relevant to their own lived experiences.

At this juncture it is important to keep in mind that this focus area is meant to generate a listing of candidate phenomena that will serve as a resource for task design. The listing is not set in stone; rather, it is to be revisited and updated as you continue with the unpacking and move through the subsequent *NGSA* steps. For this reason, do not worry if your listing of phenomena is a brainstorm that is incomplete or only represents broad possibilities. There are a number of sources that can support your efforts in identifying candidate phenomena: students' interests and experiences, everyday phenomena from local communities, science topics in the news or on television, science articles in journals, other teachers, textbooks, and the internet. An *NGSS*-specific resource for examples of phenomena that relate to DCIs is the book *Disciplinary Core Ideas: Reshaping Teaching and Learning* (Duncan et al., 2017).

DCI Focus Area 5: Identifying Student Strengths and Challenges

The aim of this DCI focus area is to outline anticipated student strengths and challenges with the DCI elements. To guide our query into strengths we ask, "What everyday experiences have students had that would be relevant for this DCI?" This question encourages us to consider the assets that students can bring to bear for making sense of the DCI elements. These assets might include community knowledge and experiences, cultural events, and/or other everyday experiences that intersect with learning science. Documenting strengths has the value of informing designers about assets that could be used to make tasks that connect more directly with students' lives, are more personally relevant and engaging, and are more strongly anchored in the surrounding real world.

When identifying challenges, we ask, "What commonly held ideas differ in important ways from the scientifically accepted understanding of this DCI?" The aim here is to document likely conceptual challenges that students will face as they engage in sensemaking with the DCI elements. Awareness of challenges helps task designers in two ways. First, it can inform the types of supports or scaffolds that may need to be designed into tasks to support students in using and applying the sub-ideas of the DCI elements. Second, knowing the commonly held ideas can enable designers to create features in tasks—such as strategically worded prompts—that will better pinpoint what students know and can do in relation to the target science ideas. For example, research on students' understanding of conservation of mass has shown that many middle school students believe that the total mass decreases in a chemical reaction when a gas is produced (Özmen, 2004) or when a solid is dissolved in a liquid (Özmen, 2004; Stavy, 1990). Accordingly, a task that is designed around a chemical reaction phenomenon might include a prompt to pinpoint whether students are indeed holding on to this common idea that can interfere with making sense of conservation of mass. Insight regarding students' ideas can be drawn from research, such as scholarly research on student ideas in science (e.g., Driver, Guesne & Tiberghien, 1985) and research on learning progressions (e.g., Mohan, Chen & Anderson, 2009), as well as from one's own teaching experiences. NSTA resources include a wide range of books and articles on students' ideas in science.

Unpacking Science and Engineering Practices

When unpacking the SEPs, an important aim is to identify and define the aspects of performance that comprise each practice and determine the appropriate ways students will engage in the practice at a particular grade or grade band. By *aspect*, we mean a necessary part of an SEP that students would need to perform anytime the SEP was put into use. For example, the practice of modeling can involve students in developing a model and/or using a model. An essential aspect of developing a model includes representing the appropriate and necessary elements of a system or phenomenon to help predict the *how* or explain the *why*. Another essential aspect of developing a model includes representing the relationships or interactions among the model elements. Beyond identifying the aspects, we also need to consider *how* students will be expected to engage with the SEP. Consider again the practice of modeling. Students in the elementary grades are expected to engage with this practice in ways different from middle school and high school students. For example, in grades K–2, students are expected to develop and use models that take the form of diagrams or physical replicas to represent concrete events or design solutions. In the upper elementary grades, the modeling practice expands to include evaluating and revising simple models to explain or predict phenomena. The expectations become more sophisticated in the secondary grades when students develop, use, and revise more complex models to explain, test, and predict phenomena. These grade band differences are important to document so that the assessment tasks that are developed include the SEP aspects that are appropriate for students and calibrated to their respective grade bands. Another important aim for unpacking the SEPs is to identify the productive intersections between the target SEPs in a bundle and other complementary SEPs and then describe how they relate to one another. The value in doing this is that it enables you to earmark a set of SEPs that could potentially be used across a range of tasks for assessing progress toward achieving the PE bundle. Similar to instruction that might engage students with multiple SEPs as they build proficiency with a PE bundle, assessment should also consider the role of other complementary SEPs for assessing students' developing proficiencies.

These two important aims are among five that we try to accomplish when unpacking an SEP. Organized into five focus areas, the SEP unpacking process involves (1) describing the SEP and its essential aspects of performance, (2) identifying productive intersections between the SEP and other SEPs that are relevant for building toward the PE or PE bundle, (3) specifying the evidence that will be needed to demonstrate the SEP, (4) describing students' prerequisite knowledge, and (5) identifying student strengths and challenges. Table 3.3 presents the five focus areas for SEP unpacking along with guiding questions for each area.

TABLE 3.3. Focus Areas and Guiding Questions for SEP Unpacking

SEP Focus Area	Guiding Questions
1. Describing SEP Aspects	• What is a clear grade-appropriate definition of the SEP? • What are the essential aspects of the SEP that students are to perform?
2. Identifying intersections with other SEPs	• What are the productive intersections between this SEP and other SEPs? • What related SEPs could students engage with in concert with the DCI and CCC of the PE?
3. Specifying evidence to demonstrate the SEP	• What is the evidence required for students to demonstrate a high level of proficiency with the SEP?
4. Describing Prerequisite Knowledge	• What knowledge and capabilities do students need to demonstrate the SEP?
5. Identifying Student Strengths and Challenges	• What everyday ways of knowing and doing relate to this SEP and how are they similar and different from the SEP? • What common challenges might students encounter as they are developing sophistication in their use of the SEP?

SEP Focus Area 1: Describing the SEP and Its Essential Aspects of Performance

The SEP unpacking involves identifying the essential aspects of the SEP and describing how students will be expected to perform the practice. We have found that crafting a clear, grade band appropriate description of the target SEP is a productive starting point for this focus area. This description should include a definition of the practice (e.g., What is the practice of developing and using models?) along with its value for science and science learning (e.g., Why is this practice important?). Once the practice itself is clarified, then its essential aspects should be identified and described. The essential aspects of any SEP are the "must-dos" for performing that practice. Importantly, these are described with the student in mind (e.g., What must a student do to perform this practice?). For example, when using a model to explain or predict a phenomenon, you might articulate that students should always use the key elements of the model to explain how and why the phenomenon occurs. Another essential aspect might be that students should always express how the model relates to the actual phenomenon. Yet another might be that students should always note one or more limitations of the model.

For any practice, we always get our footing by first referring to the *Framework* (NRC, 2012) and then broadening our reach to include a range of resource articles on the practice. Importantly, we strive to get clarity on the practice and what it should look like for students to use the practice in the classroom. Helpful resources include the *NGSS*'s Appendix F on Science and Engineering Practices and Schwarz and colleagues' (2017) book, *Helping Students Make Sense of the World Using Next Generation Science and Engineering Practices*. Both of these resources describe how K–12 students can be expected to engage with the SEPs in *NGSS* classrooms.

SEP Focus Area 2: Identifying Productive Intersections Between the SEP and Other SEPs

In this focus area, we identify complementary SEPs that could productively work together along with the DCI elements from the PE bundle. For example, if modeling is the anchor practice of a PE bundle, we would identify other related SEPs that might intersect productively in the context of the PE bundle. One intersection might involve the practice of data analysis where models can be developed based on results of data analysis. Another intersection might be found with using models to form coherent explanations about phenomena. And yet another productive intersection might relate modeling with the practice of communicating information. Note that the aim is to identify intersections that are relevant for building toward the PE bundle. The value of taking the time to identify these intersections is that they can also be considered for use in assessment tasks. Importantly, any that are identified in this focus area could be included in the upcoming step of mapping the dimensions. We have found the NSTA book on science and engineering practices (Schwarz, Passmore & Reiser, 2017) to be a helpful resource for this focus area.

SEP Focus Area 3: Specifying the Evidence That Will Be Needed to Demonstrate the SEP

In this focus area, we specify the evidence required for students to demonstrate a high level of proficiency with the target practice at the appropriate grade band. As a rule, we make this specification for each essential aspect of the practice that is identified from SEP focus area 1. To illustrate, for the practice of modeling we refer back to all the essential aspects for using a model to explain or predict a phenomenon. For each, we note the evidence that students will need to provide at a proficient level of performance. One aspect of using a model is being able to identify the key elements of that model. The evidence of performance might be that students identify the appropriate and necessary elements in the model that are needed to explain the phenomenon under study. Another aspect of using a model is describing the relationship between the model and the phenomenon. The evidence of performance might be that students correctly describe how the model's elements relate to the real-world phenomenon.

In Table 3.4, we list five essential aspects for using models to explain or predict phenomena. For each one, we briefly describe the aspect and the evidence that a middle school student would need to demonstrate. Specifying this evidence is of high value for both task design and rubric design. In the case of task design, the performance specifications make clear what tasks will need to elicit for a given SEP. For rubric design, the performance specifications pinpoint the range of evidence that students will need to provide to demonstrate proficiency with the SEP. Importantly, this helps strengthen the alignment between tasks and rubrics. Note that for this focus area we have found that it is tremendously helpful to pull from resources that are aimed at your target grade band. Fortunately, there is an increasing pool of resources, including NSTA journal articles and books on the eight practices across the elementary and secondary grade bands.

TABLE 3.4. Aspects of the Modeling Practice and the Performance Expected of a Middle School Student When Using Models

Essential Aspects of "Modeling"	What a Student Should Be Able To Do When "Using Models"	How Will a Student Demonstrate Proficiency?
Explanation/ prediction	Ability to generate an **explanation** or make a **prediction** using a model.	Student uses the model to correctly **explain or predict** a real-world phenomenon.
Parts of the model	Ability to **identify** the parts or attributes of a model.	Student correctly **identifies** the appropriate and necessary parts or attributes in a given model.
Relationships among model parts	Ability to **describe** the relationships and interactions among parts of a model.	Student correctly **describes** the appropriate relationships and/or interactions among the parts in a given model.
Relationship between model and real-world phenomenon	Ability to **describe** the relationship of parts of a model to their real-world counterparts or available data.	Student correctly **describes** the relationship between the model and a real-world phenomenon by describing the relationship between model parts and the real-world phenomenon or available data.
Limitations	Ability to **identify** the limitations of a model.	Student correctly **identifies** the limitations of a model given its intended purpose.

SEP Focus Area 4: Describing Students' Prerequisite Knowledge

When unpacking SEPs, it is important to note the knowledge and capabilities that students will be expected to draw upon for performing the practices. This includes what you can expect students to already know from prior grades and instructional experiences—knowledge and capabilities that are relevant for further developing proficiency with the target SEP. The value of identifying and describing relevant prerequisite knowledge is that it makes clear the knowledge that students will need to employ but that will not need to be assessed. This information, in turn, can help designers calibrate their assessment tasks so that the tasks measure what is most appropriate to measure. For instance, by the middle grades, students should already be familiar with the idea that models can be developed and used to help explain or predict phenomena. Accordingly, a task that uses modeling might ask middle school students to evaluate and revise a model to better meet the goal of explaining or predicting. In this example, the task leverages students' prerequisite knowledge about model use to engage them in a more sophisticated aspect of the practice.

Appendix F of the *NGSS* provides a matrix for the progression of SEPs across grades K–12. The matrix highlights ways that students engage in the SEPs at each grade band. This trajectory is useful for understanding the minimum level of proficiency expected of students at the target grade band and how proficiency at the target grade band builds on these previous experiences with the practice.

SEP Focus Area 5: Identifying Student Strengths and Challenges

In this focus area, we attend to the diverse ways that students make sense of the world and how students' everyday sensemaking practices (i.e., their ways of knowing and doing) can relate to a given SEP. This includes thoughtful attention to the language and literacy components of the practice and how they relate to students' own language and literacy repertoires. Identifying these connections enables designers to recognize and create opportunities within tasks to engage students with the SEPs in ways that connect with students' own experiences, perspectives, and ideas.

Engaging in the SEPs is language-intensive, and it is important to recognize that students bring a wide range of literacy skills to using the various practices. Students' language and literacy repertoires can be leveraged effectively in assessment tasks when designers are aware of and accepting of students' modes of language use. It is at this juncture, too, when it is important to consider how to support those who are learning English as they learn science in being able to demonstrate their knowledge and capabilities with the practices. For example, though it might be necessary to use formal language and technical terms in demonstrating a practice, designers can allow for other instances where informal language and more familiar sentence structures can be used.

Another focus is on identifying common challenges that students might encounter as they are developing sophistication in their use of the SEPs. For instance, students in the early grades often view models as replicas of objects rather than having explanatory or predictive power. In the middle grades, students often struggle to develop and use models to represent phenomena that are too small or too large to observe directly. Awareness of challenges—such as those just mentioned with regard to modeling and with other SEPs including their essential aspects—can help task designers in two ways. First, it can inform the types of supports or scaffolds that may need to be designed into tasks to assist students in demonstrating a practice. For instance, students who are developing proficiency with a given practice may benefit from a prompt within a task that reminds them to use one or more essential aspects of that practice. Second, it can identify opportunities for assessing the extent to which students have overcome a common challenge. Both are important considerations for task design. Resources for identifying strengths and challenges of students using the practices include Schwarz and colleagues' NSTA book on science and engineering practices (2017), NSTA journal articles on the practices (e.g., Bell et al., 2012; Damelin et al., 2017; Forbes et al., 2015), and contemporary research on the practices (e.g., Berland & McNeill, 2010; Metz, 2004; Schwarz et al., 2009).

Unpacking Crosscutting Concepts

The CCCs are conceptual resources that can be used across all science disciplines by students for making sense of phenomena and unfamiliar problems. When unpacking the CCCs, it is important to consider the cross-disciplinary nature of them as well as how students can engage with the CCCs in the context of a given PE bundle. It is also important to identify and describe the aspects of the CCC that are grade band appropriate. In the *NGSA* design process, the CCC unpacking is conducted across five focus areas: (1) describing the CCC and the essential aspects that students are expected to apply, (2) identifying productive intersections between the CCC and other CCCs, as well as the SEPs identified in the SEP unpacking, (3) specifying the evidence that will be needed for applying the CCC, (4) describing students' prerequisite knowledge, and (5) identifying students' strengths and challenges. Table 3.5 presents the five focus areas for unpacking a CCC along with guiding questions for each area.

TABLE 3.5. Focus Areas and Guiding Questions for CCC Unpacking

CCC Focus Area	Guiding Questions
1. Describing CCC Aspects	• What is a clear grade-appropriate definition of the CCC? • What are the essential aspects of the CCC that students are to apply in the context of the PE or PE bundle?
2. Identifying Productive Intersections	• How does this CCC intersect with the SEP and within the set of DCI elements? • What are the productive intersections between this CCC and the other SEPs identified in the SEP unpacking? • What other CCCs could students engage with in concert with the DCI and SEP of the PE or PE bundle?
3. Specifying Evidence for Applying the CCC	• What is the evidence required for students to demonstrate a high level of proficiency with the CCC?
4. Describing Prerequisite Knowledge	• What knowledge and capabilities do students need to demonstrate the CCC?
5. Identifying Student Strengths and Challenges	• What everyday experiences have students had that would be relevant for this CCC? • What common challenges might students encounter as they are developing sophistication in their use of the CCC?

CCC Focus Area 1: Describing the CCC and Its Essential Aspects

This focus area entails defining the CCC and describing its essential aspects at the target grade band. We have found it worthwhile to first define the CCC broadly, with attention paid to how the CCC serves as a conceptual resource that can be applied across disciplines. Then we describe the essential aspects related both to the target grade band and the PE bundle. It is important to consider, too, any information that might further serve to highlight how the CCC is intended to be used. For example, especially noteworthy is that matter and energy are unique in that they are considered both as DCIs and a CCC. It will be helpful to differentiate them so that you have a clear definition of energy and matter as a CCC. A major difference is that the CCC focuses primarily on conservation of matter and energy, whereas the DCIs and their elements focus on mechanisms of change involving matter and energy (Anderson, Nordine & Welch, 2021).

Because each CCC increases in complexity and sophistication across the grades, it will be most advantageous to determine the essential aspects of the CCCs according to grade band. The CCC Patterns, for example, involves multiple types, such as repeated occurrences, similarities and differences (which often includes comparing and categorizing), and correlations and trends. Their nature and use will vary across the grades. Table 3.6 shows some essential aspects for pattern use involving similarities and differences for grades 6–8. Chapter 4 of the *Framework* (NRC, 2012), Appendix G of the *NGSS* (NGSS Lead States, 2013), NSTA's *Atlas of the Three Dimensions* (Willard, 2020), and *Crosscutting Concepts: Strengthening Science and Engineering Learning* (Nordine & Lee, 2021) are helpful resources for getting started.

CCC Focus Area 2: Identifying Productive Intersections Between the CCCs and Between CCCs and SEPs

Similar to our approach when identifying intersections between SEPs (see SEP focus area 2), we also pay attention to the relationships between the target CCC and other CCCs. We identify commonalities among CCCs that could productively work together within the context of the PE bundle. Note that you can find many possible intersections between the CCCs, but we recommend that you aim to identify only those that seem most productive for the PE bundle. For instance, there are numerous potential meaningful pairings between the CCC of Patterns and other CCCs. Broadly, patterns can be used to identify cause-and-effect relationships (CCC of *cause and effect: mechanism and explanation*) and patterns can help determine how a system is changing over time (CCC of *stability and change*), to name just a couple of the uses of patterns in relation to the other CCCs. By giving attention to the interplay of the target CCC within the PE bundle, you can more easily narrow down to the most productive intersections.

In this focus area we also consider the productive intersections between the target CCC and the SEPs. When intentionally put to use along with relevant SEPs, CCCs provide a vantage point for sensemaking and problem-solving. For example, patterns can be sources of evidence and also serve as evidence in arguments; cause-and-effect relationships can be represented in models to predict phenomena; and systems can be observed at various scales using models. Viewed in this way, CCCs are most meaningful when used within the context of practices. In turn, CCC–SEP pairings are most meaningful when used within a disciplinary context for sensemaking and problem-solving. For this reason, we strive to identify pairings that will work best in light of the DCI components and their elements within the PE bundle. For example, the target CCC of Patterns and the practice of Analyzing and Interpreting Data are the *NGSS* pairings in MS-PS1-2 from our example PE bundle, and are intended to work in concert with the DCI components and their elements for chemical reactions. In MS-PS1-2, students "analyze and interpret data on the properties of substances before and after the substances interact to determine if a chemical reaction has occurred." Another productive intersection for the CCC of patterns that works in tandem with the DCI components and elements of the PE can be found with the practice of constructing explanations. Patterns can be sources of evidence for explanations, and explanations can account for patterns in data. In the case of this PE, the patterns in data can be used to construct an explanation that will account for how substances have physical and chemical characteristics that change during chemical reactions. Resources for this focus area include Appendix G of the *NGSS*, Chapter 4 of the *Framework*, and the NSTA Press book, *Crosscutting Concepts: Strengthening Science and Engineering Learning* (Nordine & Lee, 2021).

CCC Focus Area 3: Specifying the Evidence That Will Be Needed for Applying the CCC
In this focus area, we specify the evidence required for students to demonstrate proficiency with the target CCC at the appropriate grade band. We make this specification for each essential aspect of the practice that is identified from CCC focus area 1. The evidence is written at a proficient level of performance, meaning that each statement describes the successful grade-appropriate demonstration of the essential aspect. To illustrate, for the CCC of Patterns we refer to the pattern types and their essential aspects. For each type, we note the evidence that students will need to provide for the essential aspects. In middle school, students use a range of pattern types such as repeating occurrences (e.g., cyclical patterns), correlations and trends (e.g., relationships among variables), and similarities and differences (e.g., classifications). To complete this focus area, we would write statements of evidence for what students will demonstrate when they use each of these types of patterns. Table 3.6 illustrates how evidence for the essential aspects of one type of pattern,

similarities and differences, might be documented at the middle school level. In addition to using *Crosscutting Concepts: Strengthening Science and Engineering Learning* (Nordine & Lee, 2021), we recommend referring to the article, "The second dimension—crosscutting concepts: Understanding A Framework for K–12 Science Education" (Duschl, 2012) and Appendix G of the *NGSS*.

TABLE 3.6. Aspects of the Crosscutting Concept of Patterns and the Performance Expected of a Middle School Student When Using "Similarities and Differences in Patterns"

Essential Aspects of "Similarities and Differences"	What a Student Should Be Able To Do When Using "Similarities and Differences in Patterns"	How Will a Student Demonstrate Proficiency?
Determining similarities and differences	Ability to **determine** similarities or differences among two or more quantities or properties by making comparisons.	Student makes a correct **determination** about similarities or differences among two or more quantities or properties in a model, phenomenon, or data.
Characterizing amount or degree of similarity or difference	Ability to **characterize** the amount or degree of difference among two or more quantities or properties.	Student correctly **characterizes** the amount or degree of difference among two or more quantities or properties.
Categorizing based on similarities and differences	Ability to **categorize** based on similarities and differences among objects or entities.	Student correctly **categorizes** objects or entities according to their similarities and differences within a model, phenomenon, or data.
Explaining based on similarities and differences	Ability to **explain why** quantities or properties are similar or different. Ability to **explain why** categories based on patterns are meaningful.	Student correctly **explains why** quantities or properties are similar or different. Student correctly **explains why** categories based on patterns are meaningful given their intended purpose.

CCC Focus Area 4: Describing Students' Prerequisite Knowledge

Our effort in this focus area is on documenting what students are already expected to know about the CCCs from instructional experiences at prior grade levels. Notably, attention is paid to the knowledge and capabilities that are relevant for further developing proficiency with the target CCC. The CCCs are emphasized in different ways across the K–12 grades, and it will be important to both identify the prerequisite knowledge for demonstrating a given CCC and the extent to which students have had the opportunity to engage with it. For example, the crosscutting concept of Energy and Matter is addressed differently over time. Attention during the elementary years is centered on building ideas about matter, and much less on energy; in middle and high school, students learn to use energy ideas more explicitly to make sense of an increasing variety of phenomena and systems. Another example comes from the crosscutting concept of Patterns, where students are likely to have more experience with the pattern type of similarities and differences in the elementary grades than with correlations and trends. For determining prerequisite knowledge, the progression matrix in Appendix G: Crosscutting Concepts in the *NGSS* and *The NSTA Atlas of the Three Dimensions* (Willard, 2020) are resources where you can trace the prerequisites from every grade. Another helpful resource can be the scope and sequence from *NGSS*-designed curriculum materials from earlier grades.

CCC Focus Area 5: Identifying Student Strengths and Challenges

In our final focus area, we direct attention to students' everyday sensemaking as it relates to the CCCs and identify both assets and common challenges. All students have prior first-hand experiences and ideas about the crosscutting concepts from their everyday interactions with the world around them (Goggins et al., 2021) and it will be important to identify those that are salient to the target CCC. For instance, as students navigate their everyday world they notice patterns, encounter different sizes and scales, and find themselves in situations where they try to figure out what causes natural occurrences. These are valuable assets that designers can draw upon to sketch scenarios and create opportunities within tasks to connect with students' own experiences, perspectives, and ideas. When this is accomplished, students benefit as well. For instance, using everyday language related to the CCCs within a relatable task scenario could serve as an entry point for students to get their bearings for making sense of a novel phenomenon in an assessment task.

When identifying common challenges, it can also be helpful to consider students' everyday experiences and how language use and intuitive ideas can potentially be a tripping point. For the CCC of Energy and Matter, for instance, students across all ages experience phenomena every day in which energy seems to be "used up" and disappear (Anderson, Nordine & Welch, 2021). Yet, an important idea regarding energy and matter conservation is that both entities must always be somewhere. They cannot spontaneously appear or disappear since they cannot be created nor destroyed; rather, they move through phenomena and systems. Awareness of challenges can present opportunities for assessing the extent to which students have overcome them. Moreover, knowing the challenges can inform the sketching of relatable scenarios and problem situations for students to demonstrate their developing knowledge of how to use the CCCs. Finally, this information can help designers to write prompts that explicitly address CCC aspects (e.g., *How do the patterns in the data support the conclusion?*).

A valuable resource for identifying strengths and challenges of students using the crosscutting concepts is the book *Crosscutting Concepts: Strengthening Science and Engineering Learning* (Nordine & Lee, 2021). This NSTA book also includes many useful references for exploring the CCCs and the emerging research literature on students' ideas and performance related to them.

Guidance for Mapping the NGSS Dimensions

To design assessment tasks that reflect the integrated nature of the dimensions, the design process must synthesize the three deconstructed dimensions from the unpacking back together in ways that promote integrated proficiency with the PEs. Through *integrated dimension maps,* we can express key relationships between the DCI elements elaborated in the unpacking and then identify how aspects of the SEPs and CCCs also elaborated in the unpacking can work with these disciplinary relationships to promote students' integrated proficiency. Integrated dimension maps are visual representations that show the relationships between DCI elements and their sub-ideas, and how all three dimensions relate to one another within the context of a PE bundle. These maps are of high value because they provide a global view of the broad terrain of the PE bundle and inform the creation of intermediary three-dimensional performances that we call *learning performances*. Each learning performance takes on the three-dimensional structure of a PE and covers part of the multidimensional terrain of a PE bundle. We use the integrated dimension maps to set the boundaries for our learning performances, which ultimately serve as keystones in the process for designing assessment tasks that can be used to gauge students' progress toward achieving the PEs (see Chapter 4). The mapping step is guided by the question, *How do the unpacked dimensions relate to one another?* The mapping entails (1) laying out the DCI terrain to show how the sub-ideas derived from the DCI unpacking are connected to one another, and (2) integrating the SEPs and CCCs with the DCI sub-ideas to show how the dimensions are intended to work together for each connection that is made.

How Do We Lay Out and Make the Connections Between the DCI Sub-ideas?

Much like a typical concept map (e.g., Novak, 1990), the mapping initially takes the form of a diagram that depicts the key sub-ideas with arrows that show the relationships among them. To create the initial diagram, we lay out the sub-ideas of the DCI elements that were identified in the first focus area of DCI unpacking (*DCI focus area 1: Elaborating the DCI elements and sub-ideas*). The sub-ideas should be arranged to illustrate the important relationships and linked with arrows to make those relationships clear. Then, we refer again to the DCI unpacking to craft succinct statements that further clarify the grade-appropriate connections between the key sub-ideas. A succinct statement is crafted for each arrow in the diagram. Figure 3.1 shows a diagram for our two bundled PEs MS-PS1-2 and MS-PS1-5.

FIGURE 3.1. Disciplinary core idea terrain for PE bundle MS-PS1-2 and MS-PS1-5

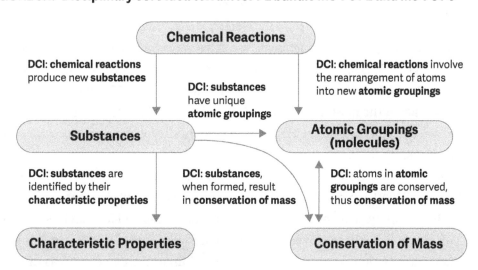

The diagram shown in Figure 3.1 lays out five sub-ideas in text boxes: substances, characteristic properties, chemical reactions, atomic groupings, and conservation of mass. The important relationships among the sub-ideas are represented by arrows with accompanying statements that describe how the sub-ideas are meant to connect. For instance, the relationship between chemical reactions and atomic groupings can be read as "*chemical reactions* involve the rearrangement of atoms into new *atomic groupings*." Note that this example diagram does not include a connection between every pair of sub-ideas; only connections that are represented in the DCI unpacking of the PE bundle are included. Accordingly, there is no connecting arrow between the ideas of *characteristic properties* and *atomic groupings* because students at the middle school level are not expected to know *why* different atomic groupings result in different characteristic properties. This relationship is not addressed until the high school grade band. The map highlights connections between the sub-ideas in a way that is not arbitrary but based on the grade-appropriate relationships between sub-ideas in the target PEs.

How Do We Integrate the SEPs and CCCs With the DCI Sub-ideas to Complete the Map?

The relationships between sub-ideas comprise the backbone of the integrated dimension map. Once the DCI terrain is mapped, we draw from both the first and second focus areas of SEP and CCC unpacking to identify SEPs and CCCs that can elicit students' proficiency for each connection made between sub-ideas. These SEPs and CCCs are oftentimes the same ones represented in the PE bundle and, accordingly, we refer to the SEP and CCC focus areas (*SEP focus area 1* and *CCC focus area 1)* to earmark essential dimensional aspects that can work for each connection. But sometimes other SEPs and CCCs are equally or more appropriate for engaging a specific disciplinary relationship expressed in the map. In these instances, we draw from *SEP focus area 2: Identifying productive intersections between the SEP and other SEPs* and *CCC focus area 2: Identifying productive intersections between the CCCs and between CCCs and the SEPs.* Note that the selection of the most appropriate SEPs and CCCs might also be informed by the scope and sequence of curriculum materials or students' level of experience with particular dimensions.

Figure 3.2 presents an example of an integrated dimension map for the PE bundle of MS-PS1-2 and MS-PS1-5. Note that each connection between sub-ideas includes a DCI relationship statement that conceptually links the sub-ideas and also includes the SEP and CCC dimensions to show knowledge-in-use. For example, the DCI relationship statement "chemical reactions produce new substances" links the two sub-ideas of chemical reactions and substances. From our SEP unpacking (focus area 2) we identified the practice of Constructing Explanations to be particularly aligned to this DCI relationship because a chemical reaction is a phenomenon that can be explained based on evidence about the starting and ending substances. For the DCI relationship linking substances and characteristic properties, we identified the practice of Analyzing and Interpreting Data because students can analyze data on characteristic properties of substances to determine whether a chemical reaction occurred. In this way, we identified two complementary SEPs that could be integrated with the disciplinary relationships between chemical reactions, the nature of substances, and characteristic properties, one of which is an anchor SEP from the PE bundle and the other which is drawn from our short list of related SEPs.

FIGURE 3.2. Integrated dimension map for bundled PEs MS-PS1-2 and MS-PS1-5

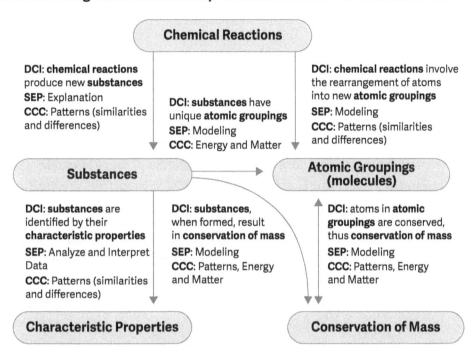

When specifying the CCCs for the linking of chemical reactions, substances, and characteristic properties, we identified patterns (notably similarities and differences) as the most salient connection to the DCI relationships and SEPs. Noteworthy is that this is also the pattern type from the PE bundle (i.e., CCC focus area 1). This is an appropriate pattern type because determining the similarities or differences (in properties or molecular structure) between the starting and ending substances in a process can be used as evidence for establishing whether a process is chemical or physical in nature. Also, the crosscutting concept of Patterns is strongly connected to both practices of Constructing Explanations and Analyzing and Interpreting Data (i.e., CCC focus area 2). Patterns, for example, can be a source of evidence for explanations; data analysis can be used to identify or characterize patterns. Once the CCC is linked with one of the SEPs and embedded in the connection of two related sub-ideas, this signifies a foremost part of the PE bundle. When all key DCI relationships integrate appropriate SEPs and CCCs, the resulting map serves as a useful representation of the full three-dimensional territory addressed by a PE bundle.

Once an integrated dimension map is created, we then use the map to articulate learning performance statements that describe multiple ways for students to demonstrate proficiency with the various relationships within the PE bundle (see Chapter 4 for guidance).

Importantly, the learning performances serve as keystones in the design of assessment tasks and rubrics (see Chapters 5 and 7 to learn how learning performances are used to construct tasks and rubrics).

Primary Takeaways

In this chapter, we pursued the question, *How can you and other teachers deconstruct and make sense of performance expectations to accurately determine what students should know and be able to do?* This question guides the foundational steps of the *NGSA* design process where we deconstruct the PEs through *unpacking* and visually represent the dimensional aspects of them through *mapping*. There are four primary takeaways from this chapter:

1. Unpacking brings to light all that is below the surface of the PEs. The unpacking exercise involves documenting the content and boundaries of the PEs and their dimensions at the appropriate grade band. Importantly, it focuses attention on the specific and meaningful aspects of the three dimensions as well as the knowledge and capabilities that students need to develop along each of them. A thorough unpacking ensures that key aspects of the dimensions are attended to in the same way each time they are used in task design. Unpacking also specifies potential intersections within and between dimensions, providing the foundation for integrating the dimensions in ways that reflect the essence of the PEs themselves.

2. When unpacking the dimensions, it is of paramount importance to always unpack with the student in mind. This includes describing students' prior knowledge and identifying likely student strengths and challenges with the dimensions; defining boundaries of what students should know and be able to do; earmarking issues of equity and inclusion that are relevant to the dimensions; identifying candidate phenomena relevant both to the PE and to students' everyday lives and interests; and sketching out possible realistic scenarios that can provide a motivating context for students to make sense of phenomena.

3. Mapping synthesizes select information from the unpacking to lay out the DCI terrain and represent the most salient intersections across the DCIs, SEPs, and CCCs. The end result of mapping is an integrated dimension map that illustrates the breadth of the PE bundle. Integrated dimension maps provide a global view of all the relationships between the sub-ideas in the PE bundle and link them with the integrated performances that students will need to demonstrate to attain proficiency.

4. Together, unpacking and mapping set the stage for constructing tasks that assess for three-dimensional learning and are suitable for classroom-based use. Organized and clearly presented information from an unpacking is an indispensable resource that will be used time and again during the *NGSA* design process. Integrated dimension maps display the various dimensional relationships and are especially useful for getting a lay of the land that encompasses the PE bundle.

Once we have completed the foundational work of unpacking and mapping, we use the integrated dimension maps in tandem with the unpacking to articulate and refine sets of intermediary performances that we call *learning performances*. A single learning performance takes on the three-dimensional structure of a PE but is smaller in scope and only partially represents the multidimensional terrain of a PE bundle. A set of learning performances collectively describe the proficiencies that students need to demonstrate to meet a PE bundle. In Chapter 4, we provide guidance on how to use the integrated dimension maps as a starting point for articulating and refining learning performances that can signal whether students are moving along a productive path to proficiency with a PE bundle.

References

Anderson, C. W., J. Nordine & M. Welch. 2021. Energy and matter: Flows, cycles, and conservation. In *Crosscutting concepts: Strengthening science and engineering learning*, ed. J. Nordine & O. Lee, 165–194. Arlington, VA: NSTA Press.

Bell, P. et al. 2012. Exploring the Science Framework: Engaging learners in scientific practices related to obtaining, evaluating, and communicating information. *Science Scope*, 36 (3), 17–22.

Berland, L. & K. McNeill. 2010. A learning progression for scientific argumentation: Understanding student work and designing supportive instructional contexts. *Science Education*. 94(5). 765–793.

Damelin, D. et al. 2017. Students making systems models: An accessible approach. *Science Scope*, 40(5), 78–82.

Driver, R., E. Guesne & A. Tiberghien. 1985. *Children's ideas in science.* Philadelphia: Open University Press.

Duncan, R. G., J. S. Krajcik & A. E. Rivet, eds. 2017. *Disciplinary core ideas: Reshaping teaching and learning.* Arlington, VA: National Science Teachers Association Press.

Duschl, R. A. 2012. The second dimension—crosscutting concepts. *Science Teacher*, 79(2), 34–38.

Forbes, C. T., et al. 2015. Scientific models help students understand the water cycle, *Science and Children*, 53(2), 42–49.

Goggins, M. et al. 2021. Broadening access to science: Crosscutting concepts as resources in the Next Generation Science Standards classroom. In *Crosscutting concepts: Strengthening science and engineering learning*, ed. J. Nordine & O. Lee. Arlington, VA: National Science Teachers Association Press.

Metz, K. E. 2004. Children's understanding of scientific inquiry: Their conceptualization of uncertainty in investigations of their own design. *Cognition and Instruction*, 22(2), 219–290.

Mohan, L., J. Chen & C. W. Anderson. 2009. Developing a multi-year learning progression for carbon cycling in socio-ecological systems. *Journal of Research in Science Teaching*, 46(6), 675–698.

National Research Council (NRC). 2012. *A framework for K–12 science education: Practices, crosscutting concepts, and core ideas*. Washington, DC: National Academies Press.

NGSS Lead States. 2013. *Next Generation Science Standards: For states, by states*. Washington, DC: National Academies Press.

NGSS Lead States. 2013. *NGSS* Appendix G: Crosscutting concepts. Washington, DC: National Academies Press.

Nordine, J. & O. Lee. 2021. *Crosscutting concepts: Strengthening science and engineering learning*. Arlington, VA: National Science Teachers Association Press.

Novak, J. D. 1990. Concept mapping: A useful tool for science education. *Journal of research in science teaching*, 27(10), 937–949.

Özmen, H. 2004. Some student misconceptions in chemistry: A literature review of chemical bonding. *Journal of Science Education and Technology*, 13(2), 147–159.

Stavy, R. 1990. Pupils' problems in understanding conservation of matter. *International Journal of Science Education*, 12(5), 501–512.

Schwarz, C. V., C. Passmore & B. J. Reiser. 2017. *Helping students make sense of the world using next generation science and engineering practices*. Arlington, VA: National Science Teachers Association Press.

Schwarz, C. et al. 2009. Developing a learning progression for scientific modeling: Making scientific modeling accessible and meaningful for learners. *Journal of Research in Science Teaching*, 46(6), 632–654.

Willard, T. (Ed.). 2014. *The NSTA quick-reference guide to the NGSS, K–12*. Arlington, VA: National Science Teachers Association Press.

Willard, T. 2020. *The NSTA atlas of the three dimensions*. Arlington, VA: National Science Teachers Association Press.

CHAPTER 4

Constructing Learning Performances That Build Toward the *NGSS* Performance Expectations

Samuel Severance, Northern Arizona University • Chanyah Dahsah, Srinakharinwirot University • Christopher J. Harris, WestEd

Perhaps the greatest challenge for meeting the vision of the *Framework* and the *NGSS* is figuring out how best to support and assess students' progress in building the proficiencies needed to achieve the *NGSS* performance expectations (PEs). The PEs are big learning goals that cannot be achieved in a single day. Thoughtfully enacted ongoing instruction is needed to support students in building the proficiencies expected of the PEs. Just as important are well-designed assessment tasks that can provide a window into students' progress in building toward them. In this way, both instruction and assessment must work together to effectively support students' three-dimensional learning over time. The science education community has made tremendous headway in developing curriculum materials to support instruction for the *NGSS*. Still, three-dimensional assessment remains a thorny challenge for those trying to gauge students' progress with building proficiencies toward a PE or PE bundle. Accordingly, you may be wrestling with a question shared by many: *How can you and other teachers use performance expectations to construct assessment tasks that can be used during instruction?*

Orchestrating instruction that engages students in three-dimensional science learning and then assessing their learning in a way that informs ongoing instruction is no simple feat. We know instruction should help students build toward the PEs so that by the end of instruction they can achieve the PEs for their grade level or grade band. We also know that PEs, each written as a single three-dimensional statement, can appear deceptively simple.

A sea captain might think that a ship can be easily maneuvered past a few floating icebergs, only to learn of a greater expanse of ice underneath the surface that the crew must worry about. Similarly, as described in Chapter 3, while many PEs seem to be straightforward and succinct statements, they are just the visible tip of the multidimensional proficiencies that students must develop over time. After unpacking the disciplinary core ideas (DCIs), science and engineering practices (SEPs), and crosscutting concepts (CCCs), it becomes clear that students' learning toward meeting a PE or a PE bundle encompasses a surprising amount. Supporting students in gaining proficiency requires having instructionally supportive assessment tasks that can serve as markers to help us see whether or not students are on a path for building toward the complex multidimensional PEs.

In this chapter, we introduce the notion of *learning performances* as a keystone in the process for designing assessment tasks that can be used—over time and in a formative manner—to gauge students' progress toward meeting complex PEs. We begin by describing what learning performances are and how they can provide guidance for creating tasks. Then, we detail how to construct learning performances and how to specify the observable evidence that students need to provide for meeting them. We end by summarizing the primary takeaways for moving from a PE bundle into a comprehensive set of smaller performances that in turn can be used to develop assessment tasks suitable for use during *NGSS* instruction.

What Are Learning Performances and Why Are They of Value?

PEs are complex and considered summative learning goals. At the elementary level, students are expected to develop proficiency across each grade; whereas at the middle and high school levels, students attain proficiency across the grade bands. Consequently, instruction should build toward a PE or a PE bundle over time, allowing students to engage with DCIs, SEPs, and CCCs in deeper and more complex ways as they make sense of phenomena and solve problems (NRC, 2012; 2015). Summative assessment tasks that align broadly to PEs are unlikely to be useful for assessing where students may need support during instruction. Instead, what are needed are three-dimensional tasks that can be used at different time points within an instructional sequence to gauge students' progress with building toward a PE or a PE bundle.

To address the goal of developing classroom-based assessment tasks that focus on three-dimensional learning, we systematically unpack *NGSS* PEs (as described in Chapter

3) and synthesize the unpacking into multiple statements that we call *learning performances*. Learning performances are intermediary performances that incorporate aspects of DCIs, SEPs, and CCCs found in PEs, but are crafted at a more specified and manageable grain-size for classroom assessment purposes. With a set of learning performances in hand, we can construct assessment tasks that can be used at formative checkpoints across a sequence of instruction. Importantly, these tasks can be used to obtain evidence of student performance in ways that cohere and build with instruction over time.

Our use of the term *learning performance* draws from the work of psychologist David Perkins (1998), specifically his notion of *performances* as opportunities for students to showcase their understanding in varied and demanding ways. From a performance perspective, understanding is not merely a matter of acquiring knowledge and being able to recall it. Instead, it is being able to put into practice what we know, oftentimes in situations where we have to think about, integrate, and apply ideas—to actually use knowledge (i.e., *knowledge-in-use*). When students use and apply what they know, these actions constitute performances that demonstrate understanding (Harris, McNeill, Lizotte, Marx & Krajcik, 2006). More recently, learning performances have been used in curriculum and assessment design to describe the variety of performances that demonstrate proficiency (e.g., Billman, Rutstein & Harris, 2021; DeBarger et al., 2015; Harris et al., 2019; Krajcik, McNeill & Reiser, 2008).

In the context of *NGSS* instruction and assessment, learning performances take the form of knowledge-in-use statements that incorporate elaborated aspects of DCIs, SEPs, and CCCs that students need to develop understanding of as they progress toward achieving PEs. A single learning performance is smaller in scope than a PE, and each learning performance describes an essential part of a PE or PE bundle that students need to demonstrate at some point during instruction to show that they are making reasonable progress toward meeting all that underlies a single PE or PE bundle. Taken together, a *set of learning performances* collectively describes the proficiencies that students need to demonstrate as they build understanding toward the PEs.

Learning performances serve as the keystone of our entire *NGSA* design process. Having a solid understanding of what learning performances are, their role in the design process, and how they are constructed is essential for creating three-dimensional assessment tasks that can be used during instruction. You might recall from Chapters 1 and 2 that our design process seeks to create an "assessment argument" which allows a teacher to obtain evidence from assessments to make a claim about a student's level of proficiency. Well, we can think of learning performances as claims about a student's proficiency toward a smaller portion of the PE. If a student meets a learning performance, then we claim that

this student is proficient in that portion of a PE. From learning performances, we can also figure out what evidence we actually need to observe to support our claims of a student's proficiency. In this way, learning performances along with the evidence we need to observe, serve as essential ingredients for developing tasks and accompanying rubrics that enable us to assess building toward performance expectations. The qualities that make for a good learning performance are summarized in Table 4.1.

TABLE 4.1. Qualities of a "Good" Learning Performance

• *Integrates* disciplinary core ideas, science & engineering practices, and crosscutting concepts
• *Identifies* an important opportunity that teachers should attend to and assess *before* the end of a unit
• *Usable* in that it provides guidance for creating tasks and rubrics that enable teachers to assess "building toward" the performance expectations
• *Functions* in relation to other learning performances to identify "what it takes" to make progress toward meeting a comprehensive performance expectation or bundle

How We Construct Learning Performances

By now, you are likely intrigued by the potential of using learning performances to create tasks and rubrics to more effectively assess students' three-dimensional learning. Let's dig into the details of how to actually construct learning performances that build toward the *NGSS* PEs. We illustrate the process using the two bundled PEs that we unpacked and mapped in Chapter 3:

- MS-PS1-2. Analyze and interpret data on the properties of substances before and after the substances interact to determine if a chemical reaction has occurred.
- MS-PS1-5. Develop and use a model to describe how the total number of atoms does not change in a chemical reaction and thus mass is conserved.

Unpacking and mapping provide the foundation for constructing learning performances. Unpacking allows us to identify the knowledge and capabilities students need to develop along each of the three dimensions to build the proficiencies associated with a PE or PE bundle. Through unpacking, we can understand the constituent parts of each of the three dimensions at a particular grade level or grade band. Mapping then allows us to bring the dimensions back together in a manner that illustrates how the dimensions are interrelated and intended to work together. Through integrated dimension maps, we can express important relationships between the DCI elements elaborated in the unpacking and identify how aspects of the SEPs and CCCs elaborated in the unpacking can work with these disciplinary relationships to promote students' integrated proficiency. Mapping synthesizes the information from unpacking to lay out the DCI terrain and visually represent the most salient intersections across the DCIs, SEPs and CCCs. An example integrated dimension map, presented in Figure 4.1, shows the multidimensional terrain of the bundled PEs MS-PS1-2 and MS-PS1-5.

FIGURE 4.1. Integrated dimension map for MS-PS1-1 and MS-PS1-2

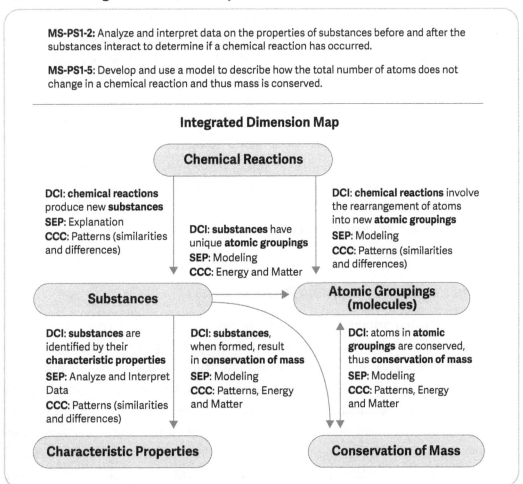

Using Integrated Dimension Mapping to Construct a Learning Performance

We use the integrated dimension maps as our starting point for articulating and refining a set of learning performances that collectively describe the proficiencies that students need to demonstrate to meet a PE bundle. We craft each learning performance as a knowledge-in-use statement that is smaller in scope, covering a designated area of an integrated dimension map and thus only partially representing aspects of the more comprehensive PE bundle. Figure 4.2 illustrates the steps we follow for specifying a single learning performance from an integrated dimension map. The mapping is a visual representation of the PE bundle that shows the relationships between the sub-ideas of the DCI elements and links them to aspects of CCCs and SEPs. In some instances, the DCI sub-ideas are linked to other closely related CCCs and SEPs as identified by the unpacking process. A single learning performance covers just part of the multidimensional terrain that resides under the surface of the PE bundle. It is intended to describe an essential part of the PE bundle that students would need to achieve at some point during instruction to ensure that they are progressing toward achieving the target PEs. Note, too, that the *learning performance takes on a three-dimensional structure similar to that of the NGSS PEs*. In the *NGSA* design process, learning performances are always written as three-dimensional performance statements that integrate DCI elements with aspects of the SEPs and CCCs.

Constructing a Set of Learning Performances

Referring back to our mapping of the PE bundle (see Figure 4.1), you can see that we can potentially create learning performances of different scopes and varying complexity by encircling different areas on the map. For example, the learning performance we specified from the encircled area in Figure 4.2 represents a relatively modest portion of the map. The same map presented in Figure 4.3 shows encircled areas representing two other learning performances. The much larger encircled area of the map shown in Figure 4.3 (b) was used to articulate a more robust learning performance: *Students develop a model of a chemical reaction that explains new substances are formed by the regrouping of atoms and that mass is conserved*. Therefore, within a given map, there can be many possible ways to demarcate areas that will shape the crafting of learning performances. Yet, it is important to remember that learning performances are intended to function together *as a set* that covers all areas of an integrated dimension map. In other words, when taken together, the learning performances we develop should not leave any area from the dimension map of

FIGURE 4.2. Constructing a learning performance from an integrated dimension map

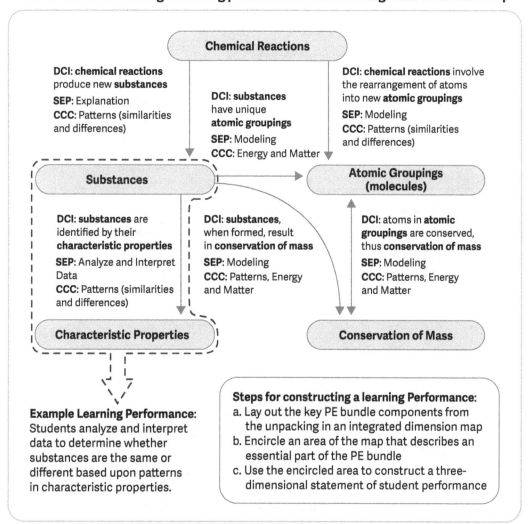

a PE or bundled PEs uncovered. Nor should any one learning performance extend outside the boundaries of the mapping. Moreover, they should collectively describe the proficiencies that students need to demonstrate to meet a PE bundle. Finally, they need to address a *range of complexity* so that they can provide evidence over time that students are building toward the PEs. Note the difference in complexity between the learning performances shown in Figures 4.2 and 4.3, indicating that instruction and assessment addressing the more comprehensive learning performance would likely come later than instruction and assessment for the learning performance in Figure 4.2 covering a smaller area of the map.

FIGURE 4.3. Delineating learning performances within an integrated dimensional map

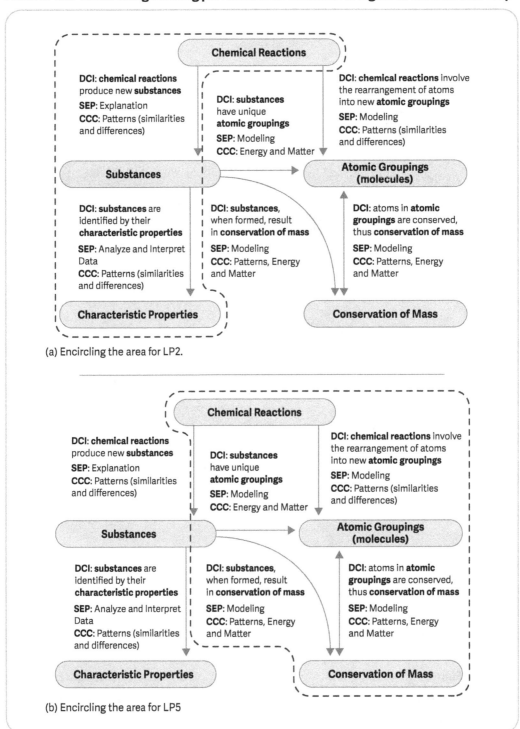

Determining the Learning Performances That Will Be Included in a Set

Given that there are many possible learning performances that can be derived from a map, how do we determine the areas of a map that should be encircled and the range of learning performances that should be included in a set? Our approach to deciding which learning performance options from the integrated dimension map should make the final cut involves examining and comparing how each learning performance addresses essential elements of the PE or PE bundle, and how these performances work together to address the PE or PE bundle as a whole. We have found that organizing all the potential learning performances from an integrated dimension map in a table can prove useful for reviewing them collectively (see Table 4.2). By reviewing the candidate learning performances individually and together, we can identify any unnecessary redundancies or overlaps of their elements. If there are, we can then discard some options until we are left with a lean set of learning performances that vary in complexity and, when taken together, serve as claims of a students' proficiency for meeting an entire PE or bundle of PEs. Another important area for review is that the set allows students to express their performances in varied ways to showcase what they know and can do. This is necessary for eventually developing a suite of tasks that will provide multiple avenues for students to demonstrate their learning related to the PEs.

TABLE 4.2. Two Bundled *NGSS* Performance Expectations With Corresponding Learning Performances

MS-PS1-2: Analyze and interpret data on the properties of substances before and after the substances interact to determine if a chemical reaction has occurred.   **MS-PS1-5:** Develop and use a model to describe how the total number of atoms does not change in a chemical reaction and thus mass is conserved.
Learning Performances (LPs) for MS-PS1-2 and MS-PS1-5
LP1: Students analyze and interpret data to determine whether substances are the same based upon patterns in characteristic properties.
LP2: Students construct a scientific explanation about whether a chemical reaction has occurred using patterns in data on properties of substances before and after the substances interact.
LP3: Students use reasoning from patterns to evaluate whether a model explains that a chemical reaction produces new substances and conserves atoms.
LP4: Students use a model to explain that in a chemical reaction the atoms are regrouped and this is why mass is conserved.
LP5: Students develop a model of a chemical reaction that explains that new substances are formed by the regrouping of atoms and that mass is conserved.
LP6: Students use reasoning about matter and energy to evaluate whether a model explains that a chemical reaction produces new substances and conserves mass because atoms are conserved.

Guidance for Writing Learning Performances as Three-Dimensional Statements

Becoming skilled in writing learning performances as knowledge-in-use statements, as seen in Table 4.2, will require practice. In keeping with the vision of the *NGSS* and its use of integrated three-dimensional science learning, we need to combine the different SEP and CCC aspects and DCI elements for a learning performance into a three-dimensional sentence structure, mirroring the structure used for the PEs in the *NGSS*. Combining the aspects and elements of the three dimensions with one another in a way that reads well and properly portrays the intended relationships often requires an iteration or two, so do not get discouraged if your first attempt looks like a bit of a word jumble! In Figure 4.4, we illustrate the three dimensions found in one learning performance. For this learning performance, we selected aspects of the science and engineering practice Developing and Using Models, disciplinary core idea elements from PS1.B: Chemical Reactions, and aspects of the crosscutting concept Energy and Matter.

FIGURE 4.4. A learning performance statement with the different dimensions highlighted

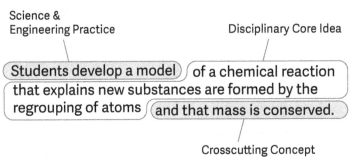

When writing a learning performance, we always strive to have all three dimensions explicit in the sentence statement. We also recommend a general format to follow when first writing a learning performance. Begin with the science and engineering practice, followed by the disciplinary core idea, and then bring in the crosscutting concept. This format provides a solid template for crafting learning performance sentence statements and, purposefully, further mirrors the three-dimensional structure of PEs. Other iterations are possible. You may sometimes find, for example, that having the crosscutting concept immediately follow the science and engineering practice leads to something more readable. Oftentimes, however, having the science and engineering practice first provides the clearest learning performance structure and immediately indicates what students need to do. Perhaps the most challenging dimension to make explicit in a learning performance is the crosscutting concept. Yet, it is critically important to make the crosscutting concept "visible" so that this dimension will not be overlooked when designing your assessment tasks (see Chapter 5).

After encircling all the areas of an integrated dimension map and sketching out all the three-dimensional statements for each area, we now have a comprehensive set of learning performances that vary in complexity (see Table 4.2). While closely related to one another, each learning performance stands on its own and tackles a different portion of the PE bundle, which is intentional. For example, the first learning performance (LP1) addresses its own essential part of the PE bundle that is distinct from the other learning performances. Taken together as a set, however, all six learning performances (LPs 1–6) address the full terrain of the bundle that is depicted in the integrated dimension map. The mountain of information encompassed in a PE bundle has now been successfully broken down into smaller, more navigable—and more easily assessed—three-dimensional statements called learning performances that can be used to develop assessment tasks suitable for *NGSS* instruction.

Describing the Integrated Proficiencies of a Learning Performance

Once a set of learning performances that builds toward a PE bundle has been written, we then describe all the integrated proficiencies that are required to demonstrate each performance. Since learning performances are the claims about what students are supposed to know and be able to do, we need to consider what students must have in their repertoires to perform them. It can be helpful to consider integrated proficiencies as the *abilities* that students will employ when doing a learning performance. It is important to take stock of the integrated proficiencies so that we clearly know the abilities that students must have to be prepared to demonstrate a learning performance.

To create an integrated proficiency for a learning performance, we refer back to the SEP aspects, DCI elements, and CCC aspects that were selected from our unpacking and mapping and used to articulate that learning performance. Take, for example, the learning performance LP5 from Table 4.2, *Students develop a model of a chemical reaction that explains that new substances are formed by the regrouping of atoms and that mass is conserved.* This learning performance includes aspects of the practice Developing and Using Models (i.e., develop a model to describe unobservable mechanisms); sub-ideas of the disciplinary core idea elements involving Chemical Reactions, Substances, and Atomic

Groupings (i.e., two elements from the component idea PS1.B: Chemical Reactions); and crosscutting concept aspects related to Energy and Matter (i.e., conservation of mass). The dimensional aspects and elements used to create LP5 are shown in Table 4.3. We use the dimensional aspects and elements to describe all the abilities that are required for demonstrating the learning performance. As in LP5, a single learning performance usually includes multiple integrated proficiencies.

When delineating the proficiencies, we use all the SEP and CCC aspects and DCI elements that are found in the learning performance. We bring the different aspects and elements together to ensure that they represent the integrated abilities for doing the learning performance. When constructing LP5, we identified three integrated proficiencies:

1. Ability to construct a model of a chemical reaction that shows regrouping and conservation of mass.

2. Ability to describe how the model shows that atoms regroup during a chemical reaction.

3. Ability to support the model by explaining that chemical reactions conserve atoms and therefore conserve mass.

Describing the integrated proficiencies points us in the direction for considering the kinds of three-dimensional evidence that we will look for to determine whether students are able to use and apply their knowledge in the manner called for by the learning performance.

TABLE 4.3. The Dimensional Aspects and Elements Used to Create LP5

SEP (3 Aspects)	DCI (2 Elements)	CCC (2 Aspects)
Develop a model to describe unobservable mechanisms: 1. Model Elements: Specify elements of the model (and their attributes) and describe why these elements are necessary. 2. Relationship Among Elements: Represent the relationships or interactions among model elements and describe why these relationships are important. 3. Correspondence: Represent the correspondence between model elements and the target phenomenon or available data.	**PS1.B. Chemical Reactions:** 1. In a chemical process, the atoms that make up the original substances are regrouped into different molecules. 2. In chemical reactions, the total number of each type of atom is conserved, and thus the mass does not change.	**Energy and Matter (Conservation):** Matter is conserved because atoms are conserved in physical and chemical processes. 1. Determine that mass is conserved. 2. Describe how mass is conserved in a particular system.

Determining the Evidence for Meeting a Learning Performance

Learning performances represent a set of claims about the knowledge students need to develop, use, and apply. Integrated proficiencies describe the abilities that students are expected to employ when doing a learning performance. Once we know the abilities that students must have to demonstrate a learning performance, we can then identify the evidence that is required for meeting each learning performance. Importantly, we aim to specify the *observable* evidence that students will need to provide to support a claim about proficiency with a given learning performance. We use the abilities that are described in the integrated proficiencies to determine the evidence and we specify this evidence in a format that we call *evidence statements*. Evidence statements are written for each of the integrated proficiencies.

What Are Evidence Statements?

Evidence statements provide direct guidance for what we should look for in a student's performance. They describe what "counts" as important when students use and apply knowledge to demonstrate a learning performance. Accordingly, the evidence drawn from a set of integrated proficiencies can be used to guide the writing of the tasks and rubrics that will be used to assess whether students have met a learning performance. For example, well-written evidence statements will denote the information required in a student's response to defend a claim about that student's overall proficiency with a learning performance. With this information in hand, a designer can intentionally create an assessment task to elicit a response from students that will provide observable data that demonstrates their level of proficiency with that same learning performance.

As we will see in Chapters 5 and 7, having evidence statements clearly specified is of high value because they allow for a more systematic and principled approach to designing aligned assessment tasks and rubrics. Without evidence statements, we would not be able to determine whether or not students have met a learning performance. Notably, the writing of evidence statements moves us from the "unpacking" stage of our *NGSA* design process into the next stage, toward where we will focus on developing tasks and rubrics.

Guidance for Writing Evidence Statements

From the first step through to the final step in our *NGSA* design approach, we strive to maintain the integration of the three dimensions of science proficiency. Unpacking and mapping enables us to move from large three-dimensional PE bundles to sets of three-dimensional learning performances that are smaller in scope. Given that evidence statements are intended to denote what constitutes meeting a learning performance, they should also follow a three-dimensional structure. Evidence statements should be in the form of clearly written expressions of three-dimensional evidence. Importantly, they should be written from the integrated proficiencies that in turn, were derived from the dimensional aspects and elements of a given learning performance. Referring back to Table 4.3, we integrated aspects of a science and engineering practice (Developing and Using Models) with elements of a disciplinary core idea topic (PS1.B Chemical Reactions) and aspects of a crosscutting concept (Energy and Matter: Flows, Cycles, and Conservation) to create a learning performance (LP5) with integrated proficiencies and evidence statements. Note that each dimension for LP5 has two or more aspects or elements that were identified from the unpacking and integrated dimension mapping of the PE bundle. For example, there are two conservation aspects of the crosscutting concept of Energy and Matter that are part of the learning performance. The science and engineering practice focuses on developing a model and includes three aspects: (1) identifying model elements, (2) describing the relationship among the elements, and (3) representing the correspondence between model elements and the real-world phenomena. There are two elements of the disciplinary core idea of chemical reactions that address the regrouping of atoms and that the number of each type of atom remains the same.

As you can see in Table 4.4, the integrated proficiencies for LP5 closely align with the learning performance as well as the structure and content of the evidence statements. In this way, the integrated proficiencies serve as a bridge connecting the learning performance to its evidence statements. For example, the evidence statement *Students' description of a model shows that atoms are rearranged during a chemical process* indicates what must be evident in the ability as described by the integrated proficiency. In turn, the ability described by the integrated proficiency is required for demonstrating the learning performance.

TABLE 4.4. Integrated Proficiencies and Evidence Statements for LP5

Learning Performance	Integrated Proficiencies	Evidence Statements
LP5: Students develop a model of a chemical reaction that explains new substances are formed by the regrouping of atoms and that mass is conserved.	Ability to construct a model of a chemical reaction that shows regrouping and conservation of mass	Students construct a model that shows atoms are correctly regrouped from reactants to products and each type and number of atoms is conserved.
	Ability to describe how the model shows that atoms regroup during a chemical reaction	Students' description of the model shows that atoms are rearranged during a chemical process.
	Ability to support the model by explaining that chemical reactions conserve atoms and therefore conserve mass	Students' statements about the model show that mass is conserved by providing evidence of number of each type of atom before and after a reaction.

Important to note is that learning performances within the same PE bundle could potentially share some overlap with one another on an integrated dimension map. For instance, learning performances LP1 (Figure 4.2) and LP2 (Figure 4.3) cover some of the same terrain on the map. Learning performances LP4 and LP5 overlap with each other too. Accordingly, on occasion when there is overlap, it is possible that a single evidence statement could correspond to more than one learning performance. This is the case for LP4 and LP5 in this PE bundle.

Using Learning Performance Expectations and Evidence Statements

So far, we have examined what learning performances are, how to delineate them within an integrated dimension map, how to write them as three-dimensional performances, and what counts as evidence of a student having met them. With evidence statements in hand, we can now use the learning performances as stepping stones, supporting students in building toward a PE in a coherent fashion. The evidence statements specify the observable evidence we need to take into account to develop tasks that are strongly aligned to each learning performance in a PE bundle (see Chapter 5). Instead of administering a single summative assessment, typically given at the end of instruction to measure proficiency

with a PE or PE bundle, we can administer a suite of tasks over time—for example, one task for each learning performance each week—making it much easier to follow students' progress over the course of instruction and as they build toward the PEs. Notably, we can also use the evidence statements for each learning performance to develop rubrics to analyze students' responses to tasks (see Chapter 7). In this way, learning performances and evidence statements give us the footing to build a coherent and instructionally supportive assessment system for understanding students' progress in meeting PEs over time.

Developing an Assessment Argument

Remember that developing assessments is a form of argumentation (see Chapter 2). Within this context, the idea behind an "assessment argument" is to make a defensible claim of a student's proficiency in meeting a complex PE or PE bundle. If using a broad and isolated assessment task, we would be hard-pressed to make a defensible claim about what aspects of a PE a student has effectively built an understanding of and what aspects they need to work on. By aligning our assessment approach to learning performances that matter for instruction, however, we can make more precise claims tied to the finer aspects of a PE (e.g., by identifying a student who has met all the aspects or elements in one learning performance but has not yet met all the aspects or elements in a second learning performance). Most importantly, we can now use evidence statements to point to precise evidence that indicates where a student is in terms of achieving a learning performance. With a precise and reliable assessment argument, we have the ability to better adjust instruction to meet students' needs.

Primary Takeaways

PEs are complex and considered summative learning goals, and therefore need to be learned over time and through a sequence of carefully designed lessons and units. In this chapter, we introduced learning performances as a pivotal step in the process for designing assessment tasks that can be used—over time and in a formative manner—to determine where students are in their progress toward meeting the complex PEs. Summative assessment tasks that align only broadly to PEs are not likely to be useful for assessing where students may need support during instruction. Instead, what are needed are three-dimensional tasks that can be used at different time points during instruction to gauge students' progress in building toward a PE bundle. There are four important takeaways from this chapter:

1. PEs represent the end goals of instruction and therefore are not immediately helpful for constructing tasks that measure students' progress toward them. Our solution is to break down a PE or PE bundle into a comprehensive set of smaller learning performances that in turn can be used to develop assessment tasks suitable for *NGSS* instruction.

2. Learning performances are knowledge-in-use statements that incorporate *aspects* of disciplinary core ideas, science and engineering practices, and crosscutting concepts that students need to develop understanding of as they progress toward achieving PEs. A single learning performance is smaller in scope and partially represents a PE. Each learning performance describes an essential part of a PE that students would need to achieve at some point during instruction to ensure that they are progressing toward achieving the more comprehensive PE. Taken together, they collectively describe the proficiencies that students need to demonstrate to meet a comprehensive PE or PE bundle.

3. We use integrated dimension maps in tandem with unpacking to articulate and refine sets of learning performances. Both the mapping and unpacking are essential to the principled articulation of three-dimensional learning performances that coherently represent the target PEs. A well-designed set of learning performances will allow students to express their performances in varied ways to showcase what they know and can do.

4. Integrated proficiencies and evidence statements work together to describe the observable evidence that students need to provide for meeting a given learning performance. The evidence drawn from a set of integrated proficiencies can be used to guide the writing of the tasks and rubrics that will be used to assess whether students have met a learning performance.

With a better understanding of how to develop a set of learning performances for a PE or PE bundle, as well as a better understanding of how to derive integrated proficiencies and associated evidence statements for a learning performance, we are now ready to move into the actual development of tasks. In the next chapter, we explore how to specify design blueprints that guide the development of knowledge-in-use tasks. The design blueprints are used to construct assessment tasks and rubrics aligned with each learning performance. The process of specifying and using design blueprints will be framed by much of what we have explored thus far, including selecting a single learning performance and leveraging its associated integrated proficiencies and evidence statements toward task creation.

References

Billman, A. K., D. Rutstein & C. J. Harris. 2021. *Articulating a transformative approach for designing tasks that measure young learners' developing proficiencies in integrated science and literacy*. Berkeley, CA: Regents of the University of California.

DeBarger, A. H. et al. 2015. Building an assessment argument to design and use next generation science assessments in efficacy studies of curriculum interventions. *American Journal of Evaluation*, 37(2), 174–192.

Harris, C. J. et al. 2019. Designing knowledge-in-use assessments to promote deeper learning. *Educational Measurement: Issues and Practice*, 38(2), 53–67.

Harris, C. J. et al. 2006. Usable assessments for teaching science content and inquiry standards. In *Assessment in science: Practical experiences and education research*, ed. M. McMahon, P. Simmons, R. Sommers, D. Debaets & F. Crowley, 67–88. Arlington, VA: National Science Teachers Association Press.

Krajcik, J., K. L. McNeill & B. J. Reiser. 2008. Learning-goals-driven design model: Developing curriculum materials that align with national standards and incorporate project-based pedagogy. *Science Education*, 92(1), 1–32.

Mislevy, R. J. & G. D. Haertel. 2006. Implications of evidence-centered design for educational testing. *Educational Measurement: Issues and Practice*, 25(4), 6–20.

National Research Council (NRC). 2012. *A framework for K–12 science education: Practices, crosscutting concepts, and core ideas*. Washington, DC: National Academies Press.

National Research Council (NRC). 2015. *Guide to implementing the Next Generation Science Standards*. Washington, DC: National Academies Press.

Pellegrino, J. W. 2016. *21st century science assessment: The future is now*. Menlo Park, CA: SRI International.

Perkins, D. 1998. What is understanding? In *Teaching for understanding: Linking research with practice*, ed. M. S. Wiske, 39–58. San Francisco, CA: Jossey-Bass.

CHAPTER 5

Developing Assessment Tasks That Provide Evidence of Three-Dimensional Learning

Chanyah Dahsah, Srinakharinwirot University • Jane Lee, Michigan State University • Christopher J. Harris, WestEd

Just as instruction should help students build proficiency with the *NGSS* performance expectations (PEs), classroom assessment should help teachers gauge students' progress toward achieving them. For assessment to serve this purpose, tasks must do more than elicit students' grasp of disciplinary core ideas. Rather, they should require students to use and apply disciplinary core ideas in concert with science and engineering practices and crosscutting concepts to make sense of phenomena or solve problems (see, e.g., Harris, Krajcik, Pellegrino & Debarger, 2019; National Research Council, 2014). When students can use and apply the three dimensions in an integrated manner, they are demonstrating "knowledge-in-use." During instruction, the process of using and applying the dimensions is referred to as *three-dimensional learning*. Three-dimensional learning is the means through which students build proficiency with the knowledge-in-use goals of the *NGSS* PEs (Krajcik, 2015; National Research Council, 2012). In this chapter, we provide guidance for answering the question, *How can assessment tasks be designed to provide evidence of three-dimensional learning so that you and other teachers can gauge students' progress with the NGSS performance expectations?*

Within the realm of the *Framework* and the *NGSS*, three-dimensional learning is the centerpiece of science instruction. Accordingly, assessment tasks for *NGSS* classrooms need to be three-dimensional and include prompts that require students to demonstrate that they are able to use and apply disciplinary core ideas (DCIs), crosscutting concepts

(CCCs), and science and engineering practices (SEPs) (Harris et al., 2019). Assessment tasks also need to be thoughtfully designed to be accessible for a wide range of students with varying backgrounds, skills, and abilities so that they can demonstrate their three-dimensional learning (Furtak, et al., 2020). To accomplish this, assessment tasks should, for example, allow for multiple modes of expression so that students can demonstrate their proficiencies in more than one way, include grade-level-appropriate vocabulary and sentence structures so that students can appropriately make sense of the tasks, and use inclusive scenarios where the phenomena are likely to have universal relevance for students. Regarding this last point, the overarching context or situation that frames each task needs to be relevant to science and/or engineering design while also relatable to students' everyday real-world experiences or interests. Moreover, tasks should be well-aligned with three-dimensional learning goals, such as learning performances or performance expectations, so that they measure what matters. Finally, tasks need to assess what they are supposed to assess fairly for all students by accommodating for diverse cultural, linguistic, and socioeconomic backgrounds. Attending to fairness allows for making valid generalizations about all students' knowledge and abilities.

The *Next Generation Science Assessment* (*NGSA*) design process provides a systematic approach for developing a variety of tasks that fulfill the important requirements for assessment of three-dimensional learning. In the beginning steps of the *NGSA* design process as illustrated in Figure 5.1, we identify our target performance expectations and systematically unpack them (for details, refer to Chapter 3). Unpacking the dimensions of the performance expectations is the foundational step in our process in which we do a "deep dive" into all that underlies the *NGSS* PEs in order to understand their assessable components. This dive aims to elaborate the various aspects of the dimensions and also includes describing students' prior knowledge and identifying likely student challenges with the dimensions; defining boundaries of what students should know and be able to do; earmarking issues of equity and inclusion that are relevant to the dimensions; and identifying phenomena relevant both to the PE and to students' everyday lives and interests. We then use the dimension elaborations from the unpacking to create integrated dimension maps that provide visual representations of the target performance expectations. The maps express key relationships between the DCI elements elaborated in the unpacking and identify how aspects of the SEPs and CCCs also elaborated in the unpacking can work with these disciplinary relationships to promote students' integrated proficiency.

Once we have completed the foundational work of unpacking and mapping, we use the integrated dimension maps in tandem with the unpacking to articulate and refine sets of intermediary performances that we call *learning performances* (for a refresher,

FIGURE 5.1. The six steps of the *Next Generation Science Assessment* design process

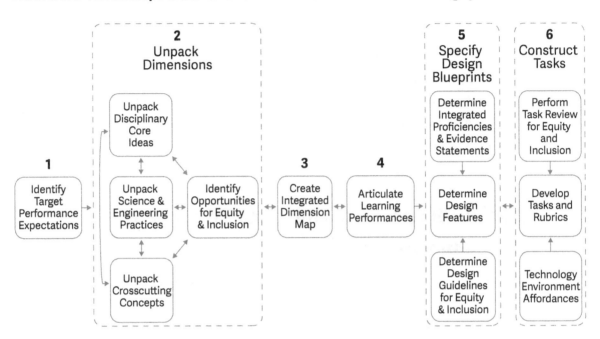

see Chapter 4). Learning performances are knowledge-in-use statements that incorporate *aspects* of disciplinary core ideas, science and engineering practices, and crosscutting concepts that students need to develop understanding of as they progress toward achieving PEs. Each learning performance takes on the three-dimensional structure of a PE but is smaller in scope and only partially represents the multidimensional "terrain" of a PE bundle. For each PE bundle, we articulate a set of learning performances that collectively describe the proficiencies that students need to demonstrate during a sequence of instruction to ensure they are progressing toward achieving the more comprehensive PE bundle. We use the integrated dimension maps to set the boundaries for our learning performances and to ensure that we cover the full terrain of the PEs.

When writing a learning performance, we also describe the integrated proficiencies that are required to demonstrate the learning performance and we identify the *observable* evidence a student needs to provide in order to show proficiency with the learning performance. We specify this evidence in a format that we call an *evidence statement*. Learning performances and their accompanying evidence statements are a keystone in our process of developing tasks that enable us to assess "building toward" the performance expectations. Importantly, they serve as the final ingredients that will be needed to develop the tasks. Once we arrive at a set of learning performances that represent a PE bundle, we use an organizational strategy called *design blueprints* to guide the principled development of tasks. A design blueprint is a document that brings together the numerous elements that must be taken into account when creating tasks for each learning performance.

In this chapter, we introduce the idea of design blueprints as an all-important organizer for using learning performances to construct assessment tasks that assess for three-dimensional learning. We begin by describing what design blueprints are and how they provide guidance for task design. Then, we detail how to specify design blueprints and use them to construct assessment tasks that align with learning performances. The chapter concludes with a summary of key points.

Design Blueprints to Guide the Principled Development of Tasks

A blueprint is a guide for making something. It is a designer's plan that details the important components that need to be taken into account during the act of creating a thing or set of similar things. The value of using blueprints is that they can make implicit design decisions *explicit* for those who will be doing the actual creating, whether the act of creating is an architectural product like a house or an assessment product like a three-dimensional task. Importantly, when designers use blueprints, they do so to ensure that the end products all meet the required design specifications. In the case of assessment, blueprints set the boundaries for what should be included or not included in tasks and they ensure that critical specifications, like task features, scaffolding levels, and format types are used consistently. Without a blueprint in hand, designers of three-dimensional tasks run the risk of important design decisions being overlooked, resulting in inconsistencies in how the dimensions are elicited and leading to unanticipated variation in what is actually assessed.

In the *NGSA* design process, we produce a design blueprint for each learning performance within a PE bundle. The blueprints are an organizational strategy that enables us to leverage all of the foundational work that is encompassed in each of the learning performances to construct assessment tasks that are aligned to them. The strategy brings together all of the specifications for task design into a single document. The specifications are organized around five key elements to ensure that the tasks are designed to elicit evidence of proficiency with a given learning performance. The first element that we include in a design blueprint is a listing of the *integrated proficiencies* that will be targeted by the assessment task. As described in Chapter 4, integrated proficiencies describe the abilities that students are expected to employ when doing a learning performance. This element provides task designers with information about what it is that students should know and be able to do in order to demonstrate a learning performance. It helps designers to answer the practical question, *What do we want students to be able to know and do within a task?* The second element that we include is the observable evidence that we should look for in a student's performance. Written as *evidence statements*, this provides designers with a clear picture of what "counts" as important when students use and apply knowledge to demonstrate a learning performance. As such, evidence statements inform both the writing of the tasks and scoring rubrics. They help designers to answer the question, *What kinds of evidence will students need to provide within a task to demonstrate proficiency with a learning performance?*

The third and fourth elements that we include in a design blueprint attend to the features of the tasks. The third element describes the features common across all the tasks—what we refer to as the *essential task features*. These can be distinguished from *variable task features*, our fourth element, that describe the attributes that can vary across tasks, such as the level of scaffolding to vary task difficulty. The fifth element is *equity and inclusion considerations* to ensure that our tasks are accessible and fair for a wide range of students with varying backgrounds, skills, and abilities so that they can demonstrate three-dimensional learning. The last three elements help designers to answer the question, *What kinds of features and formats of tasks will elicit the desired evidence from students of diverse backgrounds?*

Taken together, all five elements guide the principled development of tasks and rubrics. The first two elements provide designers with a clear picture of what inferences regarding student proficiency are to be made and they inform designers of the evidence that is required for a student to demonstrate proficiency. This steers designers toward creating tasks that align with learning performances and provide evidence as to whether students are building toward the PEs. The next two elements address task design features that enable designers to develop multiple tasks within a "family" that can share essential attributes while also varying in some features. This helps designers to create a set of tasks that vary in some useful ways yet still maintain alignment with the learning performance. The final element directs attention toward ensuring accessibility and fairness of tasks for all students. This helps designers to consider how they can reduce bias in tasks, eliminate barriers that may interfere with student sensemaking, and support engagement so that students will be more likely to persist in reading and responding to tasks.

What Are Essential and Variable Task Features?

Imagine that you are planning a new residential housing development. You have the responsibility of creating a master plan for the houses that will be built. There are certain features that will be shared across all houses that will be built in the community. Some of these features will be standard for any house, such as front doors, bathrooms, windows, sinks, and smoke detectors. And you may have some standard accessibility features, such as heights of counters, door handles, and electrical outlets that will be the same for all houses. Still, you may allow for some of the standard features to vary in some fashion, such as in the number, size, shape, and location of the standard features within each house. For instance, you might allow for the number of bathrooms to range from one to three, and you might allow for some variation in the size and shape of windows and location of doors. Yet still, there may be major features that will be optional, such as fireplaces, garages, porches, and outdoor pools. By setting the parameters for the features—those that will be standard for all houses and those that can vary in some form or by option—you can ensure that the houses in your community will have a particular look and feel. Importantly, you can ensure that those features that are essential for every house will actually be built into each house according to your specifications. In this way, and as illustrated in Figure 5.2, you can build houses that look different but function equally well for a wide range of homeowners.

We can apply this idea of planning a residential community to creating a set of assessment tasks for the learning performances in a PE bundle. When establishing the "master plan" for tasks that will align to a learning performance, we specify (1) the essential features that all tasks need to include, and (2) the important features that can vary among tasks. *Essential task features* refer to the attributes that all tasks for a particular learning performance must include. They specify the "must haves" that are described in the evidence statements of a learning performance and any other required features. For instance, if an evidence statement for a learning performance calls for students to develop a model, then all tasks designed to meet that learning performance must prompt students to create a model. Moreover, other required features identified in the unpacking or drawn from design principles should be specified. For example, a design principle for three-dimensional tasks is that every task scenario must include a situation and phenomena that is relevant to the learning performance and also of real-world relevance to students. Another design principle is that all task prompts—the directions given or questions asked—must elicit a three-dimensional response from students.

In addition to the essential features, the master plan for tasks should also specify what we call *variable task features*, or the features that can vary in some form or by option across

FIGURE 5.2. Master plan for a residential community of homes that includes both essential features and variable features

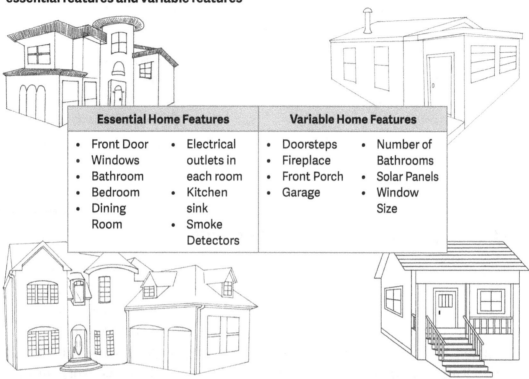

Essential Home Features		Variable Home Features	
• Front Door • Windows • Bathroom • Bedroom • Dining Room	• Electrical outlets in each room • Kitchen sink • Smoke Detectors	• Doorsteps • Fireplace • Front Porch • Garage	• Number of Bathrooms • Solar Panels • Window Size

tasks. For instance, the level of scaffolding is a feature that might be adjusted across tasks in order to reduce or increase the cognitive complexity for students. Another option that might vary is whether or not tasks include visual aids to support student comprehension. By varying a range of features like scaffolds and visual aids, we can create a set of assessment tasks that all share the same essential features and align to the same learning performance yet differ in important ways that matter for when and for what purposes you might use them. For example, two tasks designed for the same learning performance might both call for students to develop a model, but one task may be highly scaffolded to support students in carrying out the practice of modeling, whereas the other task may have no scaffolding for developing a model. In this first instance, a highly scaffolded task may be suitable for use early in an instructional sequence because the scaffolds can remind students of important components that need to be included in their models. In the second instance, a task with no scaffolding may come later in an instructional sequence and be suitable for when it will be expected that students should know the key modeling components and how to use them to develop a model to predict or describe phenomena. We provide general examples of essential task features and variable task features in Table 5.1.

TABLE 5.1. General Examples of Essential and Variable Task Features

Essential Task Features *the attributes that must be included in all tasks*	Variable Task Features *the attributes that can vary across all tasks*
Task Scenario – focuses on a real-world phenomenon that is engaging and universally relevant to students	Visual Aids – tasks can vary in the type of embedded visual aids to support comprehension of the scenario
Task Prompt – prompts must elicit a three-dimensional response	Scaffolding – the level and type of supports can vary according to task difficulty
Language Complexity – must use straightforward language that is accessible to students with diverse linguistic abilities	Modes of Response – tasks can vary in the type of response modes required (e.g., write and/or draw)

Determining Equity and Inclusion Considerations

Equity and inclusion considerations are necessary to document in a design blueprint to ensure that tasks are accessible and fair for a wide range of students. One consideration is how to provide equitable opportunities within tasks for students to make sense of the requirements for what they will be asked to do. This can be accomplished in part through thoughtful attention to grade-level-appropriate vocabulary and sentence structures so that students can properly make sense of the tasks. For example, using everyday language when describing information that is not being assessed and including clear prompts with little ambiguity can aid all students, especially those with varying reading abilities. As a general rule, tasks should be clear and comprehensible so that students with an array of linguistic abilities can access them.

Another consideration is how to support the engagement of students as they work through assessment tasks, so they will be more likely to persist in reading and responding to them. There are many ways that this can be accomplished, and we highlight two ways that we do this in our *NGSA* design work. One is to intentionally reduce bias by avoiding task elements that convey stereotypes that may interfere with student sensemaking and performance, such as stereotypes relating to gender, race, socioeconomic status, or geography. Engagement can be further supported through the use of inclusive scenarios where the phenomena are likely to have universal relevance for students. Here, for example, the overarching context or situation that frames each task needs to be relevant to science and/ or engineering design while also relatable to students' everyday real-world experiences or interests. When this is accomplished, students are better positioned to get a "footing" for engaging with the task because it leverages their experiences for using and applying knowledge to the task situation.

A final consideration is what features of tasks will enable all students to demonstrate a range of evidence of their three-dimensional learning. For example, allowing for multiple modes of expression within and across tasks can enable students to demonstrate their proficiencies in more than one way. Doing so benefits students because it can bring to light aspects of proficiency that may otherwise be overlooked if kept to just one mode. For teachers, three-dimensional learning becomes evident in ways that can deepen their own understanding and appreciation of students' developing proficiencies.

It is important to recognize that all students bring a broad range of unique experiences, background knowledge, and abilities to the classroom. By determining equity and inclusion considerations and incorporating them into design blueprints, you make explicit your attempt to attend to various student backgrounds and abilities. Importantly, you ensure that your tasks will both leverage and value students' background knowledge and experiences and connect these to rigorous science learning. These considerations include paying thoughtful attention to the language and sentence structure, including ways to reduce bias and increase inclusiveness, identifying phenomena that are relevant to students' lives, and incorporating features that enable a mixture of modes for demonstrating performance. In these ways, we deliberately strive to develop tasks that enable all students to show how they can use and apply what they know.

Specifying a Design Blueprint and Using It To Create Multiple Tasks

We use design blueprints as an organizational strategy for documenting the numerous design decisions that will guide the development of tasks. A single blueprint describes all of the design decisions for creating one or more assessment tasks that fully align to a learning performance. Importantly, every blueprint includes the five key elements described earlier—integrated proficiencies, evidence statements, essential task features, variable task features, and equity and inclusion considerations—to ensure that tasks are designed to elicit evidence of proficiency with a learning performance. When specifying a design blueprint for a learning performance, we always begin with the integrated proficiencies and evidence statements for that performance. We then use the proficiencies and statements as a starting point for identifying the essential features of tasks that will prompt students to demonstrate the learning performance. Next, we identify the features that can vary across tasks such as the level of scaffolding, scenario types, and visual aids. Together, essential and variable features enable us to create multiple tasks that differ in important ways but maintain alignment with the learning performance. We complete each design blueprint with explicit consideration for equity and inclusion, drawing from our unpacking of the PE bundle as well as guiding principles that we describe in Chapter 6.

Guidance for Specifying a Design Blueprint for a Learning Performance

We specify a design blueprint for each learning performance within a PE bundle. Take, for example, the learning performance LP5 from the set of learning performances for the bundled *NGSS* performance expectations MS-PS1-2 and MS-PS1-5 that we presented in Chapter 4: *Students develop a model of a chemical reaction that explains that new substances are formed by the regrouping of atoms and that mass is conserved.* You may recall that this learning performance includes some aspects of the science and engineering practice, Developing and Using Models, sub-ideas from disciplinary core idea elements involving Chemical Reactions, Substances, and Atomic Groupings, and crosscutting concept aspects related to Energy and Matter, notably the notion of Conservation. The design blueprint for this learning performance is shown in Table 5.2. Included at the top of the blueprint are the two bundled *NGSS* performance expectations from which the learning performance was derived. Within this blueprint, we outline three integrated proficiencies that detail the abilities that students are expected to employ when doing the learning performance LP5, along with three evidence statements that collectively describe the observable evidence that students need to provide for meeting this learning performance (refer to Chapter 4 for a description of how we derived the integrated proficiencies and evidence statements from learning performance LP5).

The evidence statements in particular are of high value because they direct the attention of designers toward the kinds of evidence that students will need to provide within a task to demonstrate proficiency with the learning performance. As such, these statements can be used to determine many of the essential features that will be designed into tasks. For example, the first evidence statement in the design blueprint, "students construct a model that shows atoms are correctly regrouped from reactants to products and each type and number of atoms is conserved," indicates several "must haves" for a well-designed and aligned task. For a student to demonstrate this evidence, the task must prompt for constructing a model of a chemical reaction that shows what is happening at the atomic level. Also, the scenario for the task should be oriented around a chemical reaction phenomenon. Another essential feature is that the task must provide a means for students to express their model of what is happening at the atomic level during a chemical reaction. This refers to how students will actually construct their model, such as through drawing on paper or by physically building ball-and-stick structures. In tandem with the evidence statements, we also revisit our unpacking of the PE bundle to be sure that we take into account pertinent information—such as clarifications, assessment boundaries, and elaborations—that may provide additional specifications for task features. Drawing from our unpacking documents, we noted that our middle school PE bundle does not expect students to know the atomic composition of substances. Accordingly, one of the essential task features informed by the unpacking is that the scenario should present information about the composition and structure of the substances involved as reactants and products in the chemical reaction. Another is that the task itself should provide a key of the atomic composition of substances for students to use in developing their models.

The second evidence statement, "students' description of a model shows that atoms are rearranged during a chemical process," indicates that a prompt (i.e., a question or cue that elicits a response) within the task is necessary for a student to know that a description is needed for how the model shows whether atoms rearrange in a chemical reaction. Without this prompt, a task would not bring about a description from the student of what the model is intended to show. For the third evidence statement, "students' statements about a model show that mass is conserved by providing evidence of number of each type of atom before and after a reaction," a prompt is also required within the task to cue the student to make a statement about how the model explains why mass is conserved during a reaction. This prompt is all-important for drawing out the ability to support one's own model.

Once we have outlined the essential task features, we turn our attention to the features that can vary in some form or by option across all tasks. When specifying variable task features, we recommend that designers consider the ways in which the scenarios or contexts that frame the tasks can vary, how the complexity of tasks can vary, and how the demands for putting to use prior knowledge and skills can vary. In our design blueprint for learning performance LP5, we indicate that scenarios across tasks can differ in three important ways—in terms of the chemical reaction phenomenon of focus, the types of situations that can contextualize the task, and the ways that comprehension can be supported for making sense of the scenario. Regarding the complexity of tasks, the blueprint provides leeway for designers on how students can express their models as well as the type and number of atoms to be included in them. The level of scaffolding for the practice of Developing and Using Models is another feature that can be adjusted in order to reduce or increase the complexity of tasks. A final variable feature is that requirements for using prior knowledge, in this case prior knowledge about chemical formulas and chemical reaction equations, can differ across tasks.

Equity and inclusion considerations round out the elements that we include in our design blueprints. For learning performance LP5, our blueprint directs designers to select a chemical reaction phenomenon that relates to students' everyday interests or

topics, and to contextualize the phenomenon in a familiar, relevant, or authentic situation. The aim here is to create a need for using and applying knowledge that arises from a thought-provoking situation that is pertinent both to the learning performance and to students' everyday lives and interests. This can promote a feeling of connectedness, support engagement with the task, and enhance motivation to work through it. Another consideration for designers is the use of written language. Within this blueprint, we recommend the use of grade level or grade band appropriate diction and sentence structures, yet also emphasize that science specific vocabulary be used when relevant to what is being assessed. Importantly, the use of straightforward everyday language can lower the cognitive demand for students so that they can better attend to what matters in the task.

TABLE 5.2. Task Design Blueprint for Learning Performance LP5 From Two Bundled *NGSS* Performance Expectations: MS-PS1-2 and MS-PS1-5

Learning Performance	LP5: Students develop a model of a chemical reaction that explains new substances are formed by the regrouping of atoms, and that mass is conserved.
Integrated Proficiencies	• Ability to construct a model of a chemical reaction that shows regrouping and conservation of mass. • Ability to describe how a model shows that atoms regroup during a chemical reaction. • Ability to support a model by explaining that chemical reactions conserve atoms and therefore conserve mass.
Evidence Statements	• Students' models show that atoms are correctly regrouped from reactants to products and each type and number of atoms is conserved. • Students' descriptions of a model show that atoms are rearranged during a chemical process. • Students' statements about a model show that mass is conserved by providing evidence of number of each type of atom before and after a reaction.
Essential Task Features	• Task presents a scenario involving a simple chemical reaction. • Task provides information about the composition and structure of the substances involved in the chemical reaction. • Task scenario provides information at the atomic levels about the reactant(s) and product(s) in a given chemical reaction. • Task prompts students to construct a model of a chemical reaction. • Task includes a key for students to use to construct a model. • Task provides a way for students to express their model of what is happening at the atomic level during a chemical reaction. • Task provides a prompt to describe how a model shows that atoms rearrange in a chemical reaction. • Task provides a prompt to make a statement about how a model explains why mass is conserved during a reaction.

Table 5.2. *(continued)*

Learning Performance	LP5: Students develop a model of a chemical reaction that explains new substances are formed by the regrouping of atoms, and that mass is conserved.
Variable Task Features	• Task scenarios can vary by phenomena and/or types of chemical reactions. • Task scenarios can vary in how information is presented (e.g., text only, text + picture, visual aids, etc.). • Tasks can require different ways to express the model, such as by drawing, writing, and/or using a modeling software tool. • Tasks can vary in the type and number of atoms to be regrouped from reactants to products in the model. • Tasks can vary in their level of scaffolding for the practice of developing and using models. • Tasks can vary in terms of increased or reduced demands for prior knowledge of chemical formulas and chemical reaction equations.
Equity and Inclusion Considerations	• Select chemical reaction phenomena that are relevant to students' lives and/or interests. • Write a reasonably compelling scenario with a familiar, relevant, or authentic situation that encourages students to engage with the phenomenon and to work through the task. • Use straightforward everyday language when describing information that is not being assessed.

Additional Pointers for Specifying Design Blueprints

The design blueprint shown in Table 5.2 illustrates one way to lay out a set of design decisions. There are, in fact, many different ways to organize design blueprints to support the efforts of task designers. We emphasize five elements that are of paramount importance, representing the core of what should be in every design blueprint. Bear in mind that there are additional elements worth considering such as assessment boundaries, common student misconceptions, known student challenges with SEPs, and task formats among others that can be included in a blueprint. Also, when specifying our own blueprints we strive to be brief so that task designers find them relatively easy to reference. Yet, you might find that more comprehensive and highly specified blueprints work better given your particular design needs. We encourage you to explore and find the right balance in terms of range of elements, level of detail, and extent of coverage or scope. Regarding this last point, be sure to include all that you deem really important for task design, but keep in mind that deep, expansive blueprints can sometimes be unwieldy for even the most experienced task designers.

How to Use a Design Blueprint to Create One or More Tasks

Clearly specified design blueprints can bring coherence and consistency to the work of creating three-dimensional tasks. They make design decisions explicit and set the boundaries for what should be included or not included in tasks. Importantly, they serve as the

ground plans for task designers, providing shared information for creating tasks aligned to learning performances and common reference points for checking work.

Design blueprints can be used to develop many possible task formats that assess three-dimensional learning. They may, for instance, be used to create comprehensive performance tasks that require extended periods of classroom time and involve instructional activities. Or they may be used to create scenario-based tasks that require just a single class session for students to complete. In our *NGSA* work, we use design blueprints to create tasks that are smaller in scope so that they are usable in the course of instruction. *NGSA* tasks are anchored in phenomena and contextualized within brief scenarios, each requiring anywhere from 5 to 10 minutes to complete, depending on the requirements of the task. The shorter task duration balances the desire to engage students in authentic science practices with the need for teachers to use the tasks flexibly during instruction and get timely information from the tasks for formative purposes. In our use of design blueprints, we strive to create tasks that require students to demonstrate various aspects of proficiency in a relatively short period of time by constructing explanations, drawing models, determining patterns in data, and by interpreting causal relationships, among other means. The tasks elicit various types of student responses including written elements, drawings, graphs, or interactions with simulations. Oftentimes, the tasks have multiple components with more than one kind of question, including short-answer questions and extended-response questions that require students to use and apply knowledge. We intend for teachers to use several tasks at appropriate points during instruction to gauge their students' progress toward achieving a given PE or PE bundle. In Chapter 8, we provide guidance on how three-dimensional assessment tasks can be used in instructionally supportive ways.

The first step in using a design blueprint is to read carefully through the elements in order to have a clear understanding of each and how they are intended to work together to define the coverage or scope of the intended tasks. This is followed by thoughtful consideration of potential phenomena plus the reasonably compelling scenarios that could contextualize the phenomena and "drive" the task. There can potentially be multiple phenomena that match with the elements in a given blueprint. An appropriate phenomenon for an assessment task will be relevant to the learning performance and also have universal relevance for students. As part of this step, it is valuable to take up any equity and inclusion considerations that can inform the identification of phenomena and the drafting of scenarios. For example, it is crucial at this juncture to consider students perspectives and experiences when choosing phenomena. It is also important that the phenomenon be presented in a situation or scenario that creates a real-world need for using and applying knowledge to make sense of the phenomenon. A third step is to give attention to the kind of prompts that can be used to elicit the SEP, DCI, and CCC aspects that have been described in the evidence statements for the relevant learning performance. This step includes reviewing the essential and variable task features that directly relate to the nature of the prompts and the modes of student response. Once a phenomenon is identified and a reasonable scenario is sketched out along with possible prompts, a series of smaller steps occurs in which attention is given to the full range of essential and variable task features along with the equity and inclusion considerations in the design blueprint. The entire process is typically carried out in an iterative manner where steps in the process are repeated as needed and comparisons are continuously made among the learning performance, elements in the blueprint, and the emerging draft task to ensure coherence between them.

When designing tasks from a blueprint, each task should fully stand alone in representing the learning performance. A well-designed task will include a phenomenon and scenario that encourages students to engage with the phenomenon and to work through the task. Responding to the task should require students to use the target DCI elements integrated with the SEP and CCC aspects of the learning performance and the prompt should be written in a manner that elicits an integrated three-dimensional response. Importantly, the prompt should relate back to the phenomenon and be consequential to the scenario so that there is a coherent narrative for students to follow. A well-designed task will also be written clearly and with expectations made explicit so that all students can grasp the intent of the task and work toward the goal of responding to it. Finally, the task should require an appropriate amount of conceptual challenge for students so that it allows for drawing classroom relevant conclusions about student performance. The foremost attributes of a well-designed three-dimensional task are summarized in Table 5.3.

TABLE 5.3. Foremost Attributes of a Well-Designed Three-Dimensional Task

Focus Area	Attributes
Phenomenon and Scenario	• Task is anchored in a phenomenon and driven by a scenario or situation that encourages students to engage with the phenomenon and to work through the task.
	• An appropriate phenomenon is consistent with and worthwhile to the learning performance, has universal relevance for students, and is addressable by using the knowledge and capabilities that are required at the grade level or grade band.
Task Prompt	• Prompt(s) relates back to the phenomenon and is consequential to the scenario so that there is a coherent narrative for students to follow and respond to.
	• Prompt(s) written in a manner that elicits an integrated three-dimensional response.

(Continued)

Table 5.3. *(continued)*

Focus Area	Attributes
Coherence	• Task creates a need for using and applying knowledge that arises from a thought-provoking situation that is pertinent both to the learning performance and to students' everyday lives and interests.
	• Task is written clearly, with scaffolds and visual aids as appropriate, and with expectations made explicit so that all students can grasp the intent of the task and work toward the goal of responding to it.
Conceptual Challenge	• Task requires students to use the target DCI elements integrated with the SEP and CCC aspects of the learning performance.
	• Task requires an appropriate amount of conceptual challenge for students so that it allows for drawing classroom relevant conclusions about student performance.

Examples of Tasks Created from a Design Blueprint

Earlier in this chapter we introduced a design blueprint for learning performance LP5 (see Table 5.2) to illustrate what design blueprints are and make clear their key elements. You might recall that LP5 is part of a set of learning performances for the bundled *NGSS* performance expectations MS-PS1-2 and MS-PS1-5 that we described in Chapter 4 (see Chapter 4 for an overview of learning performances and guidance on how to construct them). Here we showcase two tasks that were developed using the design blueprint for LP5, which emphasizes reasoning about interactions of matter to develop a model that shows that a chemical reaction regroups and conserves atoms.

Figure 5.3 spotlights the first task, *Rosy's Battery Under Water*, which was designed around the phenomenon of water electrolysis. Water electrolysis occurs when an electric current (i.e., the movement of electrons within a substance) flows through water, resulting in a chemical change where the water molecules are decomposed into hydrogen and oxygen gases. In this task's scenario, the character, Rosy, observes gas bubbles being formed when a 9-volt battery is placed in a beaker of water. The scenario directs attention to the chemical reaction that is occurring in the water and then presents a prompt that requires students to construct a model of the reaction that shows what is happening at the atomic level. Note that the prompt is purposely framed in a manner that invites a response to Rosy's wonderment about how two gases could come from the water. This artful move by the task designer helps to maintain the narrative and create coherence between the scenario and prompt. Note, too, that the directives in the prompt align to the evidence statements from the design blueprint to ensure that the task elicits evidence of a student's three-dimensional proficiency with the learning performance. Also, it is noteworthy that the expectations for responding to the task are made explicit and that they encompass the essential features that are outlined in the design blueprint.

FIGURE 5.3. Physical science assessment task: Rosy's Battery Under Water

Rosy was holding a 9-volt battery over a beaker of water and accidently dropped it in. She observed gas bubbles coming from the terminals at the top of the battery, as shown in the illustration on the right. She wondered if the bubbles were made of the same gas.

She tested the bubbles and found that some of the bubbles were made of hydrogen gas and some were made of oxygen gas. She wondered if the two gases came from the water.

How could the two gases come from the water? Draw a model that shows the chemical reaction of water changing into hydrogen and oxygen gas. Use the key below to create your model.

Key:

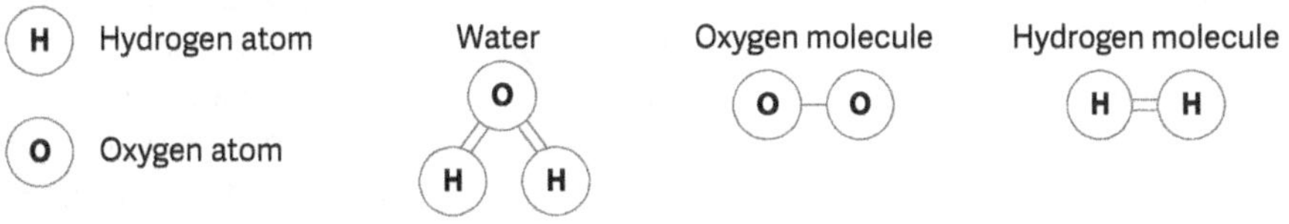

Describe how your model shows that new gases were produced when the battery was placed in the water. Based on your model, describe (1) what happened to the atoms of the water molecules during the reaction, and (2) how your model explains why mass is conserved during this reaction.

Figure 5.4 shows the second task, *Making Hydrogen Gas*, which was constructed from the same design blueprint that was used to create *Rosy's Battery Under Water* task. *Making Hydrogen Gas* was designed around the phenomenon of reacting methane gas with water to produce hydrogen gas which, in turn, can be used in hydrogen fuel cell vehicles as well as a wide range of industrial processes. Efficient, large-scale hydrogen production is a current real-world need for a green energy future. One approach to hydrogen production is to capture methane, a greenhouse gas, and react the methane molecules with water to form carbon monoxide and hydrogen. The task's scenario contextualizes the phenomenon as a focus of investigative interest for chemical engineers who are striving to produce hydrogen on a large scale for use as fuel. The scenario directs attention to the chemical reaction and puts forth a prompt that asks whether the reaction will produce a sufficient amount of hydrogen molecules. The directives in the prompt require students to use a key to construct a model of the reaction and then use the model to show how much hydrogen is produced.

CHAPTER **5**

Similar to *Rosy's Battery Under Water*, this task also aligns to the evidence statements, essential features, and the equity and inclusion considerations outlined in the design blueprint for LP5. Yet, it differs in regard to the phenomenon and scenario that drives the task, thereby creating a different motivation to figure out and use a model to explain. Another difference is found in the representation of the phenomenon, with the first task providing a picture of the water electrolysis phenomenon and the second task showing the methane reaction phenomenon as a chemical equation. Both representations support students in making sense of the task, but the representation in the second task requires that students know that a chemical equation conveys what is taking place in a chemical reaction and, as such, places an increased demand for prior science knowledge that is relevant to the grade band.

FIGURE 5.4. Physical science assessment task: Making Hydrogen Gas

Chemical engineers are working to develop fuels that do not produce carbon dioxide. Carbon dioxide is a greenhouse gas, which can cause harmful changes to Earth's climate. One fuel that chemical engineers have discussed is hydrogen gas. When hydrogen gas reacts with oxygen lots of energy is transferred to the environment, but the reaction does not produce greenhouse gases. If we had lots of readily available hydrogen gas to use as a fuel, we could reduce the amount of carbon dioxide. One way to make hydrogen gas is by reacting methane with water to produce hydrogen gas and carbon monoxide.

Methane + water → Hydrogen + carbon monoxide

Unlike carbon dioxide, carbon monoxide is not a greenhouse gas.

Does this chemical reaction produce a sufficient amount of hydrogen molecules to be used as a fuel? A model of the reaction can help answer this question. Use atoms and/or molecules from the key below to create a model that can account for all the atoms in the reaction and determine the molecules that get produced.

Key:

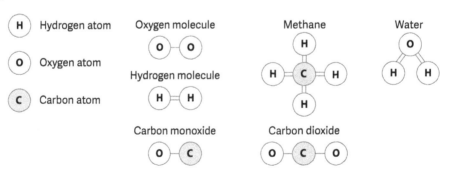

Use your model to determine how much hydrogen is produced. Based on your model, describe (1) what happens at the molecular level when methane reacts with water, (2) how your model shows that mass remains the same during the reaction, and (3) how much hydrogen is produced when one molecule of methane reacts with one molecule of water.

Guidance for Selecting Phenomena and Creating Motivating Scenarios

Assessment tasks for three-dimensional learning should be designed to be interesting and relevant to students and robust enough so that students can showcase their three-dimensional performance. A well-designed task will have as the centerpiece a phenomenon that is presented in a scenario that creates a real-world need for using and applying knowledge to make sense of the phenomenon. Importantly, a good phenomenon and surrounding scenario will set the stage for students to respond to the task by creating a context that leads students to encounter and grapple with the essential aspects of the learning performance.

An appropriate phenomenon for an assessment task will be consistent with and worthwhile to the learning performance, have universal relevance for students, and be addressable by using the knowledge, skills, and abilities that are required at the grade level or grade band. Good candidate phenomena that are worthy to include in an assessment task must contain the important science ideas of the learning performance. If a phenomenon does not stand to bring out the target dimensions, then it lacks value and is not worthwhile. Phenomena must also intersect with students' lives, reality, and/or community. Being able to relate in some fashion to a phenomenon presented in a task is important for students to engage in reasoning about that phenomenon. To achieve this for a wide range of students, a task might include an everyday phenomenon that is inherently interesting or relevant to daily experiences. Phenomena can also be meaningful when they relate to curiosities and interests about the larger natural world or relate to real-world issues that can have local or global impact. Finally, good candidate phenomena are also developmentally appropriate given students' ages, abilities, and reasoning skills. A phenomenon must be comprehensible to students and yet complex in ways appropriate for students to figure out or explain the what, how, or why of the phenomenon. If making sense of the phenomenon will require that students know, use, and apply the target knowledge at a level of complexity that is developmentally feasible within their grade band, then the phenomenon is potentially useful for assessment.

Scenarios serve to create a context for focusing student attention on aspects of phenomena that are important for responding to tasks. A well-written scenario will introduce the phenomenon and help students home in on what they need to figure out or explain. Chiefly, a good scenario will create a situation for students to grapple with the important science ideas, spark interest in figuring out the task, promote ethical scientific practice, and work in tandem with the phenomenon to provide all the pertinent information for responding to the task's prompt and directives. The situation should contextualize

the phenomena and create a need for making sense of it, thereby providing a purpose for the task. For instance, a situation might focus attention on the *how* and *why* aspects of a phenomenon that may require students to construct an explanation or use a model to explain. Another situation might problematize an aspect of a phenomenon in order to position students for planning an investigation. And yet another situation might present competing claims about a phenomenon for the purpose of requiring students to engage in argument by weighing claims and evidence. A good scenario will also aim to create interest in the phenomenon and sustain engagement with the task. This can be accomplished by crafting relatable real-world situations and characters that present the phenomena in ways likely to be meaningful and of interest to students.

Another feature of a good scenario is that it promotes ethical scientific practice and avoids depictions of harm to living things or the environment. For instance, scenarios that include descriptions of study procedures should follow standard safety guidelines, and scenarios that include living things or the environment as a focus of investigation should reflect ethical procedures. Scenarios should not depict harm or present situations where the health and welfare of livings things or the environment is endangered or portrayed in a manner that would imply a harmful outcome. As general guidance, anything that would be unacceptable in scientific practice—whether in the science classroom or in the larger scientific community—should be avoided in scenarios.

A final feature of a good scenario is that it works together with the phenomenon to create a compelling and coherent narrative that drives the task. Scenarios have the important role of creating the need for students to utilize their knowledge, skills, and abilities for making sense of the phenomenon or some important aspect of it. A task's scenario will provide all the pertinent information in a manner and sequence appropriate for enabling students to respond to the task's prompt and directives. It is important to stress that this is where attention should be given to grade-level appropriate vocabulary and sentence structure so that students with an array of linguistic abilities can properly grasp the scenario and be on good footing for responding to the task. Clearly written and accessible text can lower the cognitive demand for all students so that they can better attend to demonstrating their three-dimensional learning.

Guidance for Writing Three-Dimensional Prompts

Once the stage has been set with a phenomenon that is contextualized in a reasonably compelling scenario, the next important area of focus is writing the prompt. Prompts encompass the questions and directives that will provide evidence that students can apply their integrated learning to make sense of phenomena. There are multiple ways in which students might demonstrate three-dimensional performance and thus there are many ways that prompts can be written. That said, there are several key features that all high-quality prompts will share. The first key feature is that the prompt will require students to use the three dimensions of the learning performance. Specifically, the prompt should bring forth the SEP, DCI, and CCC aspects and elements that have been described in the evidence statements for the learning performance. This is important for ensuring that the task measures what it is intended to measure. Moreover, the prompt's format should allow for a mode of response that matches the learning performance. For instance, if a learning performance calls for students to construct a model, then the prompt should provide a response format that enables drawings or diagrams. The second key feature is that the prompt will relate back to the phenomenon and be consequential to the task's scenario so that it maintains the narrative for students to follow. This strengthens the overall comprehensibility of the task and guards against unintended and off-the-mark interpretations of what the prompt is asking of students.

The final key feature is that the prompt will be written in a straightforward manner and with a clear directive for students to follow. Including a clear prompt with little ambiguity can aid all students, especially those with varying reading abilities. Especially important to note is that a prompt may or may not include scaffolding as a means to calibrate the difficulty of responding to the task. For instance, a prompt might break down a response into its constituent parts through the use of sub-questions to ensure that students address all the requirements for a full response. Another prompt might provide explicit directives for including the components of an SEP in a response, such as reminding students to support a counterclaim with evidence when engaging in argumentation. Alternatively, a prompt might not include any scaffolding if the intended aim of the task is for students to perform more independently or to rely more on their own knowledge, skills, and abilities as requisites for responding to the task.

The Role for Creativity in Developing Assessment Tasks

The kind of tasks that are needed for today's science classrooms are far more complex than the typical true/false, multiple choice, matching, and fill-in-the-blank items that are concerned with measuring science content knowledge and recall of science principles. Science educators in the era of the *Framework* and the *NGSS* are trying to measure the active and integrated learning that comes from instruction that engages students in using and applying the three dimensions of DCIs, CCCs, and SEPs to make sense of phenomena.

Developing the rich three-dimensional assessment tasks with all the features and considerations described in this chapter will require that you draw from the technical specifications of the blueprints as well as your own designer's skill and mindset. Design blueprints provide essential guidance but alone do not turn out innovative tasks. The designer must apply expertise and craft and be willing to go through many iterations. There is no royal road—to engage productively with a blueprint and become proficient and confident in task design takes time and practice. The designer needs to have deep integrated knowledge of the dimensions as well as current knowledge of phenomena and events in the world. As with any creative enterprise, imagination and resourcefulness are also critical.

Primary Takeaways

In this chapter, we introduced design blueprints as all-important organizers for using learning performances to construct tasks that assess for three-dimensional learning. There are four main takeaways from this chapter:

1. Design blueprints guide the principled development of tasks aligned to learning performances. They set the boundaries for what should be included or not included in tasks and they ensure that critical specifications, like task features, scaffolding levels, and format types are used consistently. Without a blueprint in hand, designers of three-dimensional tasks run the risk of important design decisions being overlooked, resulting in inconsistencies in how the dimensions are elicited and leading to unanticipated variation in what is actually assessed.

2. Design blueprints include five key elements: integrated proficiencies, evidence statements, essential task features, variable task features, and equity and inclusion considerations. Integrated proficiencies and evidence statements provide designers with a clear picture of what inferences are to be made regarding student proficiency, and they inform designers of the evidence that is required for a student to demonstrate proficiency with a learning performance. Essential task features specify the common essential features that all tasks must include. Variable task features specify the important features that can vary in some form or by option across tasks. Incorporating both types of features into design blueprints allow for the development of multiple tasks that all share the same essential features but differ in ways that matter for when and for what purposes you might use them. Equity and inclusion considerations are paramount to document in a blueprint to ensure that tasks are accessible and fair for a wide range of students.

3. One or more tasks can be created from a single blueprint, but each task should stand alone and fully represent the learning performance. Foremost, each task should include a phenomenon that is contextualized in a scenario that encourages students to engage with the phenomenon and to work through the task. Responding to the task should require students to use the target aspects of the SEP, DCI, and CCC of the learning performance and the prompt should be written in a manner that elicits an integrated three-dimensional response. Importantly, the prompt should relate back to the phenomenon and be consequential to the scenario so that there is a coherent narrative for students to follow.

4. Selecting phenomena, drawing up motivating scenarios, and writing three-dimensional prompts are among the major creative activities of the task designer. Design blueprints provide essential technical information for developing tasks, but not to be overlooked is the role of creativity in bringing everything together into a well-constructed task that will measure meaningful performance. The act of developing rich three-dimensional assessment tasks with all the features and considerations described in this chapter will require that you draw from both the technical requirements of the blueprint and your own designer's skill and mindset that sparks ideas and embraces iteration.

References

Furtak, E. et al. 2020. Emergent design heuristics for three-dimensional classroom assessments that promote equity. In *The Interdisciplinarity of the Learning Sciences, 14th International Conference of the Learning Sciences (ICLS)*, ed. M. Gresalfi & I. S. Horn, Vol 3, 1487–1494. Nashville, TN: International Society of the Learning Sciences.

Harris, C. J. et al. 2019. Designing knowledge-in-use assessments to promote deeper learning. *Educational Measurement: Issues and Practice*, 38(2), 53–67.

Krajcik, J. 2015. Three-dimensional instruction: Using a new type of teaching in the science classroom. *Science Scope*, 82(8), 50–52.

National Research Council (NRC). 2014. *Developing assessments for the Next Generation Science Standards*. Washington, DC: National Academies Press.

National Research Council (NRC). 2012. *A framework for K–12 science education: Practices, crosscutting concepts, and core ideas*. Washington, DC: National Academies Press.

CHAPTER 6

Attending to Equity and Inclusion in the Assessment Design Process

Nonye Alozie, SRI International • Krystal Madden, University of Illinois
Chicago • Daniel Damelin, Concord Consortium • Joseph Krajcik, Michigan
State University • James W. Pellegrino, University of Illinois Chicago

U.S. classrooms are diverse, with students and teachers representing many cultural, linguistic, socioeconomic, ethnic, and academic backgrounds. As educators we must be prepared to teach students from diverse backgrounds. A major aspect of doing so is by creating and using an accurate assessment of their developing science proficiencies. As the diversity of our classrooms increases, we must strive to provide quality learning opportunities and activities that are fair, equitable, and inclusive by being reflective of the diversity we bring to the classroom while taking students' diverse strengths and needs into account. We must prioritize helping all students to achieve the vision of the *Framework* (NRC, 2012) and the *NGSS* (NGSS Lead States, 2013), regardless of background.

At this point, you might be wondering, "How do I attend to the unique needs of every student I teach? And how can I create and implement science assessments that ensure that all of my students are able to demonstrate what they have learned?" In this chapter, we give purposeful attention to equity and inclusion design principles that can guide the creation of three-dimensional assessment tasks for instructional use in classrooms composed of diverse learners. Our goal is to provide guidance for the development of assessment tasks that are fair and equitable in meeting and supporting the needs of all students. Since no assessment task is "perfect" in the sense of being equally fair and accessible for the vast diversity of learners, we also consider how to implement assessment tasks in the classroom in ways that attend to cultural, linguistic, socioeconomic, ethnic, and academic variations

among students. We begin with some brief background on the importance of a focus on equity and inclusion in classroom science instruction and assessment and then describe two overarching design principles regarding access and engagement. We then discuss application of those principles in the *Next Generation Science Assessment* (*NGSA*) design model described and elaborated in Chapters 2–5. Examples are then provided that highlight application of the equity and inclusion design principles to the process of adapting and implementing tasks for use during instruction. We close by discussing some of the unique ways in which technology can be used in the design and delivery of assessment tasks to support equity and inclusion. The chapter concludes by providing takeaways to help us maintain such a focus in science instruction and assessment.

The Need to Focus on Equity and Inclusion in Assessment Design and Use

All students, regardless of their prior achievements or cultural, socioeconomic, ethnic, and linguistic backgrounds, should have opportunities to participate in science experiences that help them learn how to make sense of the world around them, to be competitive in the pursuit of STEM careers, and to meaningfully contribute to society as informed citizens. The *Framework* and the *NGSS* emphasize the need to ensure that science learning opportunities are rigorous, inclusive, and equitable for all students (*NGSS*, Appendix D). Equitable learning in science means that in instruction and assessment we (1) value and leverage learners' backgrounds, including but not limited to cultural and linguistic experiences as resources, (2) connect learners' background knowledge and experiences to science, (3) and provide supports to address the wide-ranging needs of learners to ensure that all students have the same access to science learning (NRC, 2012; Lee & Buxton, 2008; Windschitl & Calabrese Barton, 2016; Leyva, McNeill, Marshall & Guzmán, 2021; McElhaney, Baker, Kasad, Roschelle & Chillmon, 2022).

"Equity in science education requires that all students are provided with equitable opportunities to learn science and become engaged in science and engineering practices; with access to quality space, equipment, and teachers to support and motivate that learning and engagement; and adequate time spent on science. In addition, the issue of connecting to students' interests and experiences is particularly important for broadening participation in science."
(A framework for K–12 science education, NRC, 2012, p. 28)

We also factor inclusivity into our design process. Inclusion goes beyond providing access; it promotes feelings of acceptance and belonging. By inclusion, we mean working to make students feel welcomed and valued through learning and assessment opportunities that leverage their diverse identities (UNESCO, 2012; Fahd & Venkatraman, 2018). Rather than retroactively adjusting science assessment tasks to be equitable and inclusive, our approach to instructionally supportive assessment design integrates design principles that make equity and inclusivity an integral part of the design and development process at the onset of and throughout the design process. Samuel Messick, a major figure in the field of educational assessment, was very concerned about the validity of assessment tasks in terms of the conclusions we can draw from student performance. He emphasized the need for ensuring that assessments do not have elements that advantage or disadvantage certain groups (Messick, 1989). For example, a task assessing photosynthesis that includes lengthy descriptions about a context that only a small portion of the student population can relate to may be easily interpretable by some students, but is likely to distract, alienate, or confuse many students. This would be a problematic case for Messick. To help minimize such concerns, our design approach includes consideration of a range of relatable scenarios that students are already familiar with or can easily gain familiarity with through supports that are built into the task, like images, videos, and/or educative descriptions.

Acknowledging diversity is the first step in creating learning and assessment experiences that reduce barriers and help students with varying needs and capabilities feel included as they develop science proficiency. The next step is to be systematic in attending to these concerns in creating such experiences. Our assessment design approach does exactly that with the goal of reducing barriers that impede diverse learners' access to achieving three-dimensional science proficiency. Our science assessments are designed to provide accessible assessment opportunities for all students, particularly marginalized students, while still maintaining academic rigor. In the next section, we describe two major design principles to help guide the design and incorporation of instructionally supportive assessment tasks into your science instruction.

Integrating Equity and Inclusion Considerations Into Assessment Design and Use

Students come to your science classroom with a range of background knowledge and experiences, and capabilities drawn from their homes and communities (Calabrese Barton, 2001). Using our design approach, we aim to value, support, and leverage the diverse back-

grounds and needs of students from the beginning of the design process. To do so, we integrate considerations about equity and inclusion by focusing on two overarching design principles: **fostering student engagement with learning** (abbreviated as *engagement*) and **providing appropriate language supports** (abbreviated as *language supports*) (Alozie et. al., 2018; Alozie, Lundh, Yang & Parker, 2021).

Principle 1: Fostering Student Engagement With Learning

Universal Design for Learning (UDL; CAST, 2011) is an approach to the design of instruction and assessment that aims to provide all students equal opportunities to succeed, no matter how they learn. The UDL principle "multiple modes of engagement" focuses on those aspects of learning that motivate students and keep them motivated to learn. This UDL principle states that students engage in information and activities that are relevant to their interests. In our design approach, we emphasize the *why* of learning by highlighting those aspects of learning that motivate students to continue to work through a challenging task. We promote the idea that valuing, leveraging, and supporting a student's culture, language, and interests can sustain their learning and help them to identify as a competent and efficacious learner of science (NRC, 2012).

Through our ongoing work with teachers and students, we have learned that not all students find the same activities or information equally relatable or inclusive, making the need for teacher adaptation or intervention essential. For example, one of our task scenarios used the act of fasting for a special holiday as the context for describing how fat and sugar are used in the body. Through classroom implementation, we found that students interpreted fasting in various ways because some did not know what fasting was and the task did not provide enough information for students to make sense of the task. This made it difficult to assess their understanding of how fat and sugars are used in the body. We can balance this kind of disconnect during instruction by taking time to explain the different ways and reasons people manage food in their everyday lives, using fasting as an example, and relating the intake of food to how the body breaks down fats and sugars. This instructional move can help teachers leverage students' experiences with food management and support students who are not familiar with such practices. Moreover, such teaching is well aligned with accepted social and cognitive learning theories which indicate that all children's learning is dependent on their ability to connect new information to their existing knowledge (e.g., Newmann, Marks & Gamoran, 1996).

Principle 2: Providing Appropriate Language Supports

The way language is used and supported can be helpful or detrimental to how students interpret an assessment task. Since science has its own culture and language, many students require supports (e.g., science-specific vocabulary, grade-level-appropriate diction and sentence structures) to learn and participate in science (Lee & Buxton, 2012). Scientific language may be vastly different from other ways we use language; the way we speak at home with our family is most likely different from the way we talk with friends, which in turn is most likely different from the way we speak when talking about a scientific investigation. In the design of our instructionally supportive assessment tasks, we provide grade-appropriate language supports to facilitate student understanding of the multiple dimensions being assessed within each assessment task. Our task development process identifies ways to support comprehension of the task, such as providing students with multiple representations like illustrations, simulations, images, and interactive graphics that can work together to make the information in the task more accessible (CAST, 2011).

With respect to *engagement*, tasks might not provide the appropriate *language supports* for every student and you might need to adapt or intervene. For example, in a task that uses plant growth to describe how photosynthesis works, students might struggle to understand how the term *grow* applies to different organisms (e.g., humans "grow up" in size and maturity, while plants "grow" in biomass, number of parts, and morphology). You might adapt the task to include literacy strategies that can help students understand what growth means for different organisms, and then help students understand how plant growth in photosynthesis is unique from other types of growth.

Incorporating the Equity and Inclusion Design Principles Into the Assessment Design Process

Assessment tasks need to be thoughtfully designed to be accessible for a wide range of students with varying backgrounds, skills, and abilities so that they can demonstrate their three-dimensional learning. To achieve accessibility, the two overarching design principles related to equity and inclusion are integrated into our *Next Generation Science Assessment (NGSA)* design process. Figure 6.1 shows the six steps of the *NGSA* design process and calls explicit attention to three steps in the process where we integrate equity and inclusion considerations. As we proceed through the design process, we strive to recognize opportunities for equity and inclusion, respond by documenting the opportunities,

realize them in our design blueprints, and reflect on our progress toward providing equitable and inclusive assessment opportunities for all students.

FIGURE 6.1. *NGSA design process that integrates equity and inclusion design principles*

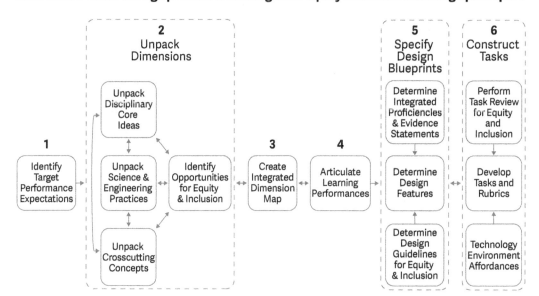

Unpack Dimensions With a Focus on Identifying Opportunities for Equity and Inclusion

In the beginning activities of the *NGSA* design process, we select our target *NGSS* performance expectations (PEs) and then systematically unpack them. Unpacking the dimensions of the PEs is the foundational step in our process in which we do a "deep dive" into all that underlies the selected PEs to understand their assessable components (for details, see Chapter 3). This dive aims to elaborate the various aspects of the dimensions and also includes recognizing and documenting the opportunities for equity and inclusion within and across the dimensions. Having chosen the PE or PE bundle you wish to target for assessment, you should begin the unpacking process by asking, "How do I identify important components of the dimensions while emphasizing equitable access for all students?" Unpacking the dimensions entails gathering information about how science and engineering practice (SEP), crosscutting concept (CCC), and disciplinary core idea (DCI) proficiencies are acquired and used in the domain. The process of unpacking also involves elaborating on *engagement* and *language supports* by identifying the different ways students can access and engage with the science phenomena and the kinds of supports that help students manage the language demands of the *NGSS* dimensions. Unpacking supports a process of merging the science being learned with student perspectives and experiences that vary due to culture, language, and socioeconomic status. For instance, when unpacking DCIs associated with photosynthesis, you can connect the science phenomena to real-world instances of students' encounters with the process and products of photosynthesis, like urban gardens, schoolyard gardens, city or state parks, or rural environments like farms.

Apply Task Design Guidelines That Support Equity and Inclusion

This step focuses on the question, "How should I design tasks to support the learning of all students?" As each learning performance is being constructed, we incorporate the equity and inclusion guidelines listed in Tables 6.1 and 6.2. Incorporating the equity and inclusion guidelines into each learning performance ensures that each task will appropriately address the diverse needs of the students.

TABLE 6.1. Design Principle *Fostering Student Engagement* With Equity and Inclusion Guidelines for Assessment Task Design

Design Principle: Fostering Student Engagement
Guideline 1: Item Engagement and Contextualization: How can the task scenario include familiar, relevant, or authentic situations? *Vivid Phenomenon-based Learning* • Build on prior knowledge and use relevant phenomena and representations (Krajcik & Shin, 2022; Lee & Grapin, 2022) • Focus on lived experiences (Windschitl & Calabrese Barton, 2016) • Connect school culture, students' home communities, and real-world contexts (Barton, 2001; Lee, 2001; Ostergaard et al., 2010) *Relating Students and Nature* • Improve student ecological awareness and fostering relations with nature (Mueller & Bentley, 2009).
Guideline 2: Cultural Sensitivity: How can the task scenario reduce bias and stereotypes for a particular gender, race, SES, or geography? • Give attention and be sensitive to issues of culture, gender, SES, race, and religion • Include context that is inclusive (i.e., context that is universal or not representative of a particular group)
Guideline 3: Visual Aids: How can the task differ in the type of visual aids to support comprehension? • Use interactive and/or simulation-based assessments and simulations as appropriate to aid comprehension (Quellmalz & Silberglitt, 2018; Grotzer et al., 2014) • Use different kinds of images and visual aids

Table 6.1 contains three guidelines that pertain to the design principle *engagement*: item engagement and contextualization, cultural sensitivity, and visual aids. Providing rich context gives us an opportunity to leverage students' funds of knowledge to bring them into disciplinary science through connections that may be easily made to their lived experiences. It is also important that our tasks are sensitive to populations that have been historically harmed through schooling and take necessary cultural sensitivity precautions to avoid additional harm. In doing so, we created sets of tasks that are inclusive of different groups to provide multiple perspectives and opportunities for diverse experiences. Finally, it is important that we consider multimodality to support student engagement. For this reason, we have used multiple representations of task components to support the different ways in which students might engage with the tasks. Providing multiple representations helps students visualize some of the content to better make connections as they learn scientific language and practices.

Similarly, Table 6.2 contains four guidelines for the *language supports* design principle: clarity of prompt, language complexity, student comprehension, and student response or expression of knowledge-in-use. As discussed earlier in the chapter in relation to the task about fasting, it is important that we provide rich context while making sure to provide adequate scaffolds and/or clearly articulating information that may be unfamiliar or confusing to some groups of students. Additionally, we must ensure that prompts are clear in the questions being posed for students to answer. This means that language must also be accessible to students and stay within grade level in terms of sentence structure, word use, etc. Scaffolding embedded within learning performance question sets has also been used to provide teachers opportunities to use different response templates based on student needs.

TABLE 6.2. Design Principle *Providing Appropriate Language Supports* With Equity and Inclusion Guidelines for Assessment Task Design

Design Principle: Providing Appropriate Language Supports
Guideline 4: Clarity of Prompt: How can the task prompts clearly elicit the student exemplary response and avoid superfluous information? • Use clear and direct language in the prompt • Make sure the prompt targets the focal knowledge, skills, and abilities of task
Guideline 5: Language Complexity: How can the task be accessible to students with limited English reading ability? • Use everyday language to describe concepts, particularly for ELL • Provide grade appropriate vocabulary words, sentence structure, and reading level • Use representations to help support vocabulary

Table 6.2. *(continued)*

Design Principle: Providing Appropriate Language Supports
Guideline 6: Student Comprehension: How can the information presented be consistent and relevant throughout the task? • Include support for prior knowledge, skills, and/or abilities • Organize information in the task to support sensemaking (e.g., labeled tables or diagrams and pictures, chunked information) • Scaffold information in the task scenario • Use terms in a consistent manner throughout the stem and prompt
Guideline 7: Student Response/ Expression of "Knowledge-in-Use": How can the task provide varied modes of student responses? • Use different response templates and/or scaffolds to support student expression of knowledge • Scaffold information in the question prompt

When designing a task, the goal is to incorporate the task features from both design principles that are most appropriate for a specific learning performance. Keep in mind that the tasks you create are meant to be instructionally supportive, so you can enhance students' learning experiences through instruction.

Perform Task Review for Equity and Inclusion

In the final stage of task development, we ask ourselves, "Do my tasks incorporate features that promote equitable and inclusive learning opportunities for a wide range of students?" To answer this question, a final review of each task is needed that follows the equity and inclusion guidelines listed in Tables 6.1 and 6.2. By using the results from a comprehensive task review, further adjustments and revisions can be made to your tasks before trying them out with students. Students will ultimately provide the most relevant feedback about how fair, equitable, and inclusive the assessments are.

Equity and Inclusion Design Principles in Action: A Task Example

The equity and inclusion design principles and guidelines help inform choices for task features that will provide information on what specifically to include when designing a task (see Chapter 5). In this section we use a life sciences task as an example to highlight how the equity and inclusion design principles have been taken into consideration and are manifest in the design of a task.

Figure 6.2 presents the three-dimensional task, *Jaden's Model*, that was developed from a design blueprint for a learning performance. You might recall from Chapters 4 and 5 that the *NGSA* design process uses learning performances for developing tasks that can be used during instruction to assess "building toward" the *NGSS* performance expectations. Learning performances are intermediary three-dimensional performance targets for instruction and assessment that can signal whether students are moving along a productive path to proficiency with a performance expectation or bundle. *Jaden's Model* was developed for a learning performance that aligns with the performance expectation MS-LS1-7: *Develop a model to describe how food is rearranged through chemical reactions forming new molecules that support growth and/or release energy as this matter moves through an organism.* This performance expectation focuses on matter and energy flow in organisms with an emphasis on the chemical transformation of food to support various life functions such as growth and movement.

FIGURE 6.2. Annotated life science assessment task: Jaden's Model

Jaden's Model was designed for the following learning performance: *Students evaluate whether a model shows that when an organism consumes food, energy is transferred to other systems within the organism and to systems outside the organism.* You might notice that the learning performance includes some of the key elements of the of the DCI and key aspects of the SEP and CCC of the overarching PE. In this task's scenario, the character, Jaden, has developed a model to show how the food transformation process unfolds for koala bears whose primary food consists of leaves from eucalyptus trees that are native to the Australian environment where koalas live. To respond to the task, students are asked to evaluate Jaden's model and do three things: (1) identify two parts of the model that show use or transfer of energy, (2) identify two parts of the model that could be improved, and (3) explain the improvements needed. For this particular learning performance we have designed multiple tasks that each address the DCI but through scenarios using different animals set in different locations and with different food sources. In Figure 6.2 we have annotated specific aspects of *Jaden's Model* to highlight aspects of the task design that illustrate attention to specific principles for supporting equity and inclusion. Table 6.3 contains a key for the specific annotations in Figure 6.2. The table entries describe particular features to which attention has been given in developing the overall task. These include creating a relatable task scenario with multiple modalities used to communicate the situation: both words and visuals; structuring the task prompt to be clear about what the student was being asked to do; and providing a visual representation of the model that the students were to reason with and about in developing their responses. Throughout, we attended to issues of vocabulary, language use, and syntax.

TABLE 6.3. Key for Interpreting Annotations in Figure 6.2

1. Unpack to identify culturally relevant phenomena and productive intersections of dimensions
a. DCI is situated in relevant and real-world context (e.g. koala bears living and eating in trees) b. Science practices and crosscutting concepts provide additional opportunities for engaging with task (e.g., modeling)
2. Specify equity in characteristic and variable task features
a. Task uses relevant and real-world context, a storyline connected to the disciplinary core idea, and supports for translating the phenomenon (e.g., a student chooses a koala to model the flow of energy through a natural system; connects students to nature) b. Task uses everyday language when describing information that is not being assessed (e.g. see underlined text) c. Task provides supports for what to include in response (e.g., take a snapshot, identify two parts of the model, circle images and text) d. Task focuses only on relevant information that will help students make sense of the phenomena (e.g., no superfluous information was added to the task) e. Task provides multiple ways to respond (e.g., write, snapshot, and draw) f. Task provides visual aids in a logical order to help students translate phenomena (e.g., image of koala living and eating in a tree)
3. Review tasks and technology for equity and fairness
a. Manipulation of model in the task promotes interaction with technology (e.g., student can revise the model by using the draw tool)

Finally, we note that another major way we have attended to issues of equity and inclusion in designing tasks for this particular learning performance is something already mentioned and which occurs early in the design process as part of the unpacking activities. Specifically, it involves considering possible situations or scenarios where the phenomenon of interest can be illustrated in ways that connect with students' lives and interests and that can engage them in helping to solve an interesting problem or explain a real-world phenomenon. While the eating behaviors of koalas may be interesting and engaging to some but not all students, teachers could instead choose to design (or select) tasks requiring students to engage in the same kind of scientific reasoning for phenomena that involve more familiar and local organisms such as squirrels in yards or forests or animals in other habitats. By having diverse possible settings for tasks and by adapting classroom implementation (see the next section) we can better approach the goals of equity and inclusion in science assessment while maintaining a focus on scientific reasoning related to core disciplinary ideas.

Implementing Assessment Tasks in Ways That Enhance Equity and Inclusion: A Classroom Example

Thus far we have described how to design instructionally supportive assessment tasks to provide equitable and inclusive access for diverse students and provided one illustration of the product of that design process. In this section, we look at an example of how a teacher enacted a three-dimensional task in her classroom, highlighting how she drew upon features of the task to further support the equity and inclusion design principles of *engagement* and *language supports* during instruction. In this case example, the teacher, Eloise, is in the midst of life science instruction focused on ecosystems with an emphasis on the core idea of how and why organisms interact with their environment and the effects of those interactions. At this juncture, her students are building toward the *NGSS* performance expectation MS-LS2-2: *Construct an explanation that predicts patterns of interactions among organisms across multiple ecosystems.* We provide this example as an illustration of how a teacher can use a three-dimensional task in an equitable and inclusive manner to support a full range of students to demonstrate their three-dimensional learning.

Figure 6.3 shows the task, *Sunflower Plant Growth*, that was created from a design blueprint for a learning performance that is part of a set of learning performances for MS-LS2-2. The learning performance, *Students analyze and interpret data to determine how organisms can interact in ways that all involved benefit or are harmed in some way and to varying degrees to meet their survival needs*, emphasizes reasoning about interactions

FIGURE 6.3. Life science assessment task: Sunflower Plant Growth

Farmer Nelson plans to plant sunflowers on her farm. She knows that sunflower height and weight is important for producing the largest flowers with the most seeds. She wonders how close she can plant the sunflowers together without affecting their growth. Farmer Nelson looked at data from studies on sunflower growth to decide the best arrangement for planting. Table 1 shows the data that she analyzed.

Table 1. Sunflower Plant Growth After 80 Days

Number of sunflower plants per 10 meters²	Sunflower Plant Growth	
	Height per plant (centimeters)	Weight per plant (Grams)
2	234	279.2
3	235	279.1
4	234	279.2
16	207	85.5
64	150	20.8
250	115	4.6
1000	100	2.1

Adapted from: Went, F. W. (1973). Competition among plants. Proceedings of the National Academy of Sciences, 70(2), 585–590.

What guidance can you give to Farmer Nelson? Use the data from Table 1 and what you know about interactions between organisms to determine the best arrangement for planting sunflowers. In your response include:

1. Data from the table to support your recommendation.

2. A description of what the data tells you about the interactions between sunflower plants when they are arranged in different ways. Be sure to tell why the number of sunflower plants planted in the same area affects their growth.

among organisms and the effects of those interactions. The aspect of the DCI associated with this learning performance and task addresses competitive interactions among organisms that affect an important survival need—growth; in this case, sunflower plants adjust to low-density population by increasing in height and weight and adjust to high-density by decreasing in height and weight. Accordingly, the arrangement of sunflower plants is important for producing the largest flowers with the most seeds. The SEP of *analyzing and interpreting data* requires students to use an investigative lens when looking at data in various forms (such as graphs, tables, and charts) and then provide an analysis and interpretation of the data. In this task's scenario, the character, Farmer Nelson, has a data table to analyze with an eye toward identifying patterns of interaction among the sunflowers. The CCC of Cause and Effect means that students will have to determine how the pattern of interactions among the same organisms may lead to changes in growth. The scenario and prompts in *Sunflower Plant Growth* require students to determine how competitive interactions between sunflower plants can cause a particular outcome, such as less growth which may affect sunflower production.

To understand how this task could be used in real-time instruction, we consider how Eloise implemented the task as part of a larger lesson. Her efforts illustrate how the embedded equity and inclusion design principles of the task can inspire instructional moves and strategies that 1) support the articulation of science proficiency and 2) engage student imagination as students demonstrate their ability to put their science knowledge to use.

Articulating Science Proficiency

To begin the lesson, Eloise wanted her students to connect their experiences to the phenomenon represented in the task. She asked herself, "What do my students already know about the phenomenon? How can I help my students relate to and talk about the phenomenon in the task?" *Sunflower Plant Growth* incorporates a real-world scenario of sunflower farming which is intended to provide a vivid phenomenon-based context for demonstrating three-dimensional learning, as articulated in Equity and Inclusion Guideline 1: Item Engagement and Contextualization. Eloise planned to leverage students' prior experiences with sunflowers to make meaningful connections to the task scenario and phenomenon. She started the lesson by involving students in a discussion to help them make connections between the *Sunflower Plant Growth* task and their experiences with sunflowers. She began by asking, "Who can tell us one thing they observe or know about the plants in this picture?" "Who has seen or grown these types of flowers?" "What else do you know about these types of flowers?" These types of questions activate relevant prior knowledge and help students relate to the scenario and phenomenon in the task. As students responded to these questions, they talked about what they already knew about sunflowers. Responses included, "They are very tall," "The flower can turn and track the sun," and "You can eat the seeds all the time." Eloise responded in turn by supporting students in finding shared experiences with others about sunflowers, prompting students to share what they knew about the growth of sunflowers, and encouraging them to share their prior experience or prior knowledge about planting and seeing sunflowers in their community. The teacher aimed to thoughtfully help students make connections with one another and also with ideas and the phenomenon. This type of discourse promotes an inclusive and equitable teaching and learning environment. Also important to note is that this discourse format can support students' articulation of their developing proficiency and help students connect and leverage their existing knowledge and experiences to their understanding of the phenomenon in the task. Eloise used the engagement and contextualization feature of the task to help students talk through their experiences and prior knowledge, thereby making connections so that they were better able to use and apply what they know and can do.

Sunflower Plant Growth requires students to formulate responses that combine information from the task scenario with data from a table. This supports student comprehension, Equity and Inclusion Guideline 6, because the task presents a coherent narrative to follow that organizes pertinent information in a table that relates to the task scenario and phenomenon. The task also aligns with Equity and Inclusion Guideline 4: Clarity of Prompt because it scaffolds the process of responding to the prompt using evidence from the data table by providing two questions that delineate the *how* and *why* of competition among the sunflowers. The teacher can scaffold the discussion by deconstructing the question prompt into smaller and more manageable components for students with different reading and English language proficiency. For example, during the discussion of the task, Eloise required that students read the entire question prompt, paraphrase it, and then relate the question to the data table. She explained, "I have to read all the information, I can't just jump to the data table." This part of the discussion not only helps students talk about how the components of the task relate to each other, but also maintains the rigor of the task while supporting their ability to respond to it.

Eloise used this as an opportunity to ask students to locate and use data in the table to help them formulate their answers, which is a key feature of the SEP Analyzing and Interpreting Data. After asking her students to identify the data in the table, she followed up with, "So not a graph. When we looked at the graph we tried to analyze it, we didn't just jump in . . . Can someone point out what we should read in the data table?" At this point of

instruction, students articulated how to gather information from the data table. Students commented that the title, columns, and headers of the data table provided information about the growth of the sunflowers.

A close look at the task reveals units of measurement that are specific to scientific language. Although this task was reviewed to have grade-appropriate reading and vocabulary, as articulated in Equity and Inclusion Guideline 5: Language Complexity, some students might be challenged with units of measurement. This is an opportunity to specifically focus on the language and vocabulary in the task. For example, while talking about how to read the data table, Eloise emphasized the units of measurement used in the table and reminded students that metric units are scientific terms, which are different from language around length and weight that they might use in everyday conversation.

As students work to respond to the task, they may struggle to navigate between the language of science, other disciplines, and everyday experience. When helping students use scientific language to articulate their understanding, it is important that you help them differentiate how certain words might differ across subject matters and contexts (Moje, Collazo, Carillo & Marx, 2001). Eloise reminded the students that science uses specific nomenclature for measurement that might differ from other subject matters to help students bridge their science learning with other encounters with and uses of measurement.

Engaging Students' Imagination During Task Implementation

The task includes an image of a field of sunflowers to support visualization and imagination that expands from the text, as described in Equity and Inclusion Guideline 3: Visual Aids. During instruction, Eloise knew that while many of her students were familiar with sunflowers and sunflower seeds, only some of her students had direct experience with growing sunflowers. She first directed their attention to the picture of sunflowers in the task and then asked them to draw flowers on the board so that students could visualize the area in which Farmer Nelson was planning to plant the sunflowers. Although a move like this does not replace what is learned through experience, it allows the students to use self-generated representations to visualize a science phenomenon in a way that supports their comprehension of the task and articulate their ideas verbally or in writing.

However, not all students may have had direct experience with farming and may struggle to imagine what closely spaced crops might look like, much less the cause and effect relationship between plant competition and access to resources. This is an opportunity to help students relate to the science in a way that is different from visualizing science phenomena through a static drawing. Role-playing has been shown to enable students to solve real-world problems in a nonthreatening learning environment while developing team-building skills in an interactive way (Bhattacharjee & Ghosh, 2013). To help students envision this kind of competition in nature, Eloise engaged the students in role-play, where she pretended to be Farmer Nelson and asked students to "plant" themselves in a confined area and pretend to be sunflowers. Eloise used role-playing to demonstrate how interactions among sunflower plants in a certain amount of space shaped their growth. She asked students to increase the population of sunflowers (more student volunteers) by using language that was familiar to them: "I need more money this holiday season from the harvest, so I need more sunflowers in there." While doing this activity, she asked students to respond to whether the spacing of the sunflowers was being "invaded" by paying attention to how crowded they felt as more people were added to the demonstration of a crowded flower garden. Eloise then drew students' attention to the data tables and asked students to use evidence from the data table to support their responses. Although role-playing does not replace the value of having a real-world experience, it can be a great way to help students visualize and physically experience phenomena (e.g., competition for resources) in a way that relates to the CCCs (e.g., *cause and effect*) when real-world experiences are not available. Role-playing activities can also help students connect what they see and experience to data presented in various forms, such as tables and charts. In Eloise's class, adding role-playing to instruction helped the students not only visualize the cause and effect relationships associated with plant competition, but also experience and experiment with how different plant densities impact plant competition.

Technology in Support of Equity and Inclusion

Our discussion has thus far avoided specific mention of the role that technology tools can play in the design or implementation of assessment tasks. Doing so was purposeful since not all teachers have access to a range of technology tools and supports for task design, task delivery, and/or for the collection of student responses. In Chapter 9 we say more about the affordances of technology with respect to multiple aspects of assessment development and use in the classroom, including broader concerns about how technology can exacerbate as well as ameliorate educational inequity. But we would be remiss if we did not briefly discuss how various assistive technologies can be used to make assessment materials more accessible to students including those with visual or physical limitations. Certain guidelines described earlier in this chapter lend themselves to facilitation through technology.

Equity and Inclusion Guideline 3: ***Visual Aids:*** *How can variation in the use of visual aids support comprehension?* Often a picture is worth more than a thousand words. Visual aids to support comprehension include images, videos, and prebuilt simulations. Images which illustrate and clarify the assessment scenario or phenomenon can be embedded along with mechanisms for enlarging the image and fading out other distracting text or visual components of the task. Videos provide a way for individuals to observe phenomena at their own pace and provide for student review as many times as necessary. Some aspects of simulations are like videos, allowing the student to "rerun" the simulation as many times as they like, while also providing a more interactive way to explore this visual aid for understanding.

Equity and Inclusion Guideline 5: Language Complexity: *How can tasks be made more accessible to students with limited English proficiency or reading ability?* Currently, most delivery of technology-enhanced assessments is done through a web browser. Modern browsers have several assistive technologies built in to support language processing. If you highlight a word, a menu can be brought up to show a definition. It is also common for a browser to be able to read aloud text that is highlighted. Services exist to translate text, such that students could copy text from the assessment task and paste it into another web-based service for translation.[1] Some browsers have translation capabilities built in as well. Additionally, one could design assessment tasks with their own custom interactive glossary containing words known to be potentially difficult for some percentage of students.

Equity and Inclusion Guideline 7: ***Student Response/Expression of "Knowledge-in-Use":*** *How can tasks provide varied modes of student responses?* Technology provides many avenues for students to express themselves. Written responses are easily collected on both paper and online forms. However, web browsers have spelling support facilitating both student expression and the teacher's ability to comprehend the student's intentions. Beyond writing, technology provides multiple ways for students to express their conceptual understanding through drawings, model diagrams, and snapshots of annotated simulations depicting some aspect of the phenomenon students are observing or investigating. By providing background drawings, custom tools, stamps, and/or a predetermined set of model components, modeling tools can help scaffold student responses.

Going Beyond the Guidelines With Assistive Technology: Many browsers or operating systems have built in assistive technologies specifically to address issues related to visual or physical impairments. These include:

---

1   Free Google translate service: translate.google.com

- Magnification: One can scale anything in a browser window to make text easier to read, images easier to interpret, or simulations easier to manipulate.
- High contrast settings: Some people benefit from amplification of the contrasts between light and dark. Many operating systems support color filters or high contrast settings for the display.
- Screen readers: Online materials are conducive to either built-in or dedicated applications that facilitate navigation and reading of text and other objects on computer screens.
- Voice dictation: Most operating systems support voice dictation into web browser text fields. This facilitates written responses when manipulating a keyboard is problematic.

Primary Takeaways

In this chapter, we explored designing and implementing instructionally supportive assessment tasks that attend to the various needs of your students. We suggest that you use the design principles *fostering student engagement with learning* and *providing appropriate language supports* as guidelines for designing your own tasks to support student demonstration of three-dimensional proficiency and when implementing tasks in diverse classrooms. The key takeaways of the chapter include:

1. Knowing your students' backgrounds and experiences can help you determine which equity and inclusion guidelines would most effectively support your students as they develop their proficiency toward *NGSS* PEs. Ask yourself, *Which groups of students in my class will be less familiar with the task and what can I do to support their learning?*

2. When designing a task for possible use, consult our Design Principles and Equity and Inclusion Guidelines to find ways to incorporate engagement and language supports throughout the task development process. Make sure to include task features from both design principles while refraining from overdesigning with too many task features.

3. Use a set of assessment tasks and various instructional resources (e.g., strategies and rubrics) during instruction to cover a wide range of supports that are built into the tasks (see Chapters 7, 8, and 9). Bundling the tasks ensures that students will encounter varying tasks with different features, increases the likelihood that

students will build connections to the science that is being assessed, and promotes three-dimensional responses from students.

4. During implementation, support students as they work to articulate their developing science proficiency by incorporating discussions that help them relate to the task features, such as the task scenario and visual representations. In addition, implement strategies that awaken students' imagination, like role-playing games, and allow students to not only visualize the task problem but also experiment with different conditions.

5. If using technology in the design or implementation of the task, use the Design Principles and Equity and Inclusion Guidelines to further increase access to the task for students. Consider how students with a diverse range of learning needs might benefit from technological enhancements to the task.

Equipped with knowledge of how to create assessment tasks that will leverage and value students' background knowledge and experiences and connect these to rigorous science learning, you stand to support three-dimensional learning more adeptly for the full range of students you teach. Our design principles for equity and inclusion highlight the value of supporting all learners and provide a systematic way to integrate these principles throughout the assessment development process. In the next chapter, we will explore how to create rubrics for three-dimensional tasks to assist you in evaluating student performance and considering implications for student feedback and instruction. As you continue, think about how rubric development can also be informed by the design principles for equity and inclusion.

References

Alozie, N. et al. 2018. *Designing and Developing NGSS-Aligned Formative Assessment Tasks to Promote Equity*. Paper presented at the annual conference of National Association for Research in Science Teaching, Atlanta, GA

Alozie, N. et al. 2021. *Designing for Diversity Part 2. The Equity and Inclusion Framework for Curriculum Design*. Rockville, MD: National Comprehensive Center at Westat.

Bhattacharjee, S. & S. Ghosh. 2013. Usefulness of role-playing teaching in construction education: A systematic review. In *49th ASC Annual International Conference, San Luis Obispo, CA*.

Calabrese Barton, A. 2001. Science education in urban settings: Seeking new ways of praxis through critical ethnography. *Journal of Research in Science Teaching*, 38(8), 899–917.

CAST. 2011. *Universal design for learning guidelines version 2.0*. Wakefield, MA: Author.

Fahd, K. & S. Venkatraman. 2019. Racial Inclusion in Education: An Australian Context. *Economies*, 7(2), 27.

Grotzer, T. A., & M. Shane Tutwiler. 2014. Simplifying causal complexity: How interactions between modes of causal induction and information availability lead to heuristic-driven reasoning. *Mind, Brain, and Education*, 8(3), 97–114.

Krajcik, J.S. & N. Shin. 2022. Project-based learning. In *Cambridge Handbook of the Learning Sciences*, ed. R. K. Sawyer, 3rd ed. New York: Cambridge.

Lee, O. 2001. Culture and language in science education: What do we know and what do we need to know?. *Journal of Research in Science Teaching*, 38(5), 499–501.

Lee, O. & C. A. Buxton. 2010. *Diversity and equity in science education: Research, policy, and practice*. Multicultural Education Series. New York, NY: Teachers College Press.

Lee, O., & S. E. Grapin. 2022. The role of phenomena and problems in science and STEM education: Traditional, contemporary, and future approaches. *Journal of research in science teaching*, 59(7), 1301–1309.

Leyva, L. A. et al. 2021. "It seems like they purposefully try to make as many kids drop": An analysis of logics and mechanisms of racial-gendered inequality in introductory mathematics instruction. *The Journal of Higher Education*, 92(5), 784–814.

McElhaney, K., et al. 2022. *A field-driven, equity-centered research agenda for OpenSciEd: Updated version* [White paper]. Digital Promise.

Messick, S. 1989. Validity. In *Educational measurement*, ed. R. L. Linn, 3rd ed. 13–103. New York, NY: American Council on Education & Macmillan.

Mueller, M. P., & M. L. Bentley. 2009. Environmental and science education in developing nations: A Ghanaian approach to renewing and revitalizing the local community and ecosystems. *The Journal of Environmental Education*, 40(4), 53–64.

National Research Council (NRC). 2012. *A framework for K–12 science education: Practices, cross-cutting concepts and core ideas*. Washington, DC: National Academies Press.

Newmann, F. M., H. M. Marks & A. Gamoran. 1996. Authentic pedagogy and student performance. *American Journal of Education*, 104(4), 280–312.

NGSS Lead States. 2013. *Next Generation Science Standards: For states, by states*. Washington, DC: National Academies Press.

Ostergaard, E., et al. 2010. Students learning agroecology: phenomenon-based education for responsible action. *Journal of Agricultural Education and Extension*, 16(1), 23–37.

Quellmalz, E. S., & M. D. Silberglitt. 2018. Affordances of science simulations for formative and summative assessment. In *Technology enhanced innovative assessment: Development, modeling, and scoring from an interdisciplinary perspective*, ed., H. Jiao & R. W. Lissitz, 71–94. Charlotte, NC: Information Age Publishing.

UNESCO. 2012. Addressing exclusion in education: A guide to assessing education systems towards more inclusive and just societies. Paris: UN. Available online: https://unesdoc.unesco.org/ark:/48223/pf0000217073 (accessed on 20 January 2019).

Windschitl, M. & A. Calabrese Barton. 2016. Rigor and equity by design: Locating a set of core teaching practices for the science education community. In *Handbook of research on teaching* 5th ed., ed. D. H. Gitomer & C. A. Bell, 1099–1158.

CHAPTER 7

Developing and Using Rubrics That Integrate the Three Dimensions of Science Proficiency

Kevin W. McElhaney*, Digital Promise • Phyllis Pennock*, Great Minds PBC • Diksha Gaur, University of Illinois Chicago • James W. Pellegrino, University of Illinois Chicago • Christopher J. Harris, WestEd

*(*co-first authors)*

Three-dimensional assessment consists of the tasks that teachers can use to draw out students' three-dimensional learning and the rubrics that can be used to interpret and evaluate students' performance on those tasks. Consider a classroom scenario where students are working through a three-dimensional task, one that assesses their science proficiency with a learning performance. In general, proficiency can be thought of as requiring a high degree of performance where a learner must use and apply knowledge in varied and demanding ways. Within the province of the *Framework* and *NGSS*, science proficiency requires students to have a high degree of performance using the three interconnected dimensions of disciplinary core ideas (DCIs), crosscutting concepts (CCCs), and science and engineering practices (SEPs) to make sense of phenomena or design solutions to problems.

Imagine in your classroom scenario that students completed the task, *Rosy's Battery Under Water*, that was introduced in Chapter 5 and is shown in Figure 7.1. To successfully respond to this task, students must use and apply their knowledge to construct a model that illustrates how two gases can be produced from water through a chemical reaction process. In their responses, students are asked to use a key to construct their models of the reaction and then describe how their models show what happened at the level of atoms and molecules. Once students have responded to the task, you are faced with a challenge shared by many teachers in *NGSS* classrooms: figuring out how to evaluate student responses on three-dimensional tasks. A *rubric* would be helpful in this scenario, one that could help you meaningfully interpret the learning demonstrated in the responses. With a well-designed rubric in hand, you can make accurate and equitable judgments about student performance, differentiate varying levels of performance among students, and get your footing for providing actionable feedback to students (Arter & McTighe, 2001). Also important is that you can use what you learn to inform your own teaching practice and facilitate next steps with instruction.

In this chapter, we provide guidance for answering the question: *How can rubrics be developed and used that will help you and other teachers to "see" students' progress in building three-dimensional proficiency?* We begin by describing how to create rubrics for tasks in which learners need to use the three dimensions to make sense of phenomena or solve problems. We then offer practical guidance for using rubrics to gain insight into student performance and pinpoint individual and collective strengths as well as areas for further growth. The chapter concludes with suggestions for how to take steps with using rubrics in innovative ways that go beyond grading and toward improved teaching and learning.

FIGURE 7.1. Physical science assessment task: Rosy's Battery Under Water

Rosy was holding a 9-volt battery over a beaker of water and accidently dropped it in. She observed gas bubbles coming from the terminals at the top of the battery, as shown in the illustration on the right. She wondered if the bubbles were made of the same gas.

She tested the bubbles and found that some of the bubbles were made of hydrogen gas and some were made of oxygen gas. She wondered if the two gases came from the water.

How could the two gases come from the water? Draw a model that shows the chemical reaction of water changing into hydrogen and oxygen gas. Use the key below to create your model.

Key:

(H) Hydrogen atom	Water	Oxygen molecule	Hydrogen molecule
(O) Oxygen atom			

Describe how your model shows that new gases were produced when the battery was placed in the water. Based on your model, describe (1) what happened to the atoms of the water molecules during the reaction, and (2) how your model explains why mass is conserved during this reaction.

Where and How Does Rubric Development Emerge From the *NGSA* Design Process?

Assessment tasks for three-dimensional learning are very different from the typical kinds of tasks that require students to only recall what they know. With tasks that call for students to use and apply the three dimensions, the expectation is that they will use the three dimensions together to make sense of a phenomenon or problem and then generate multi-dimensional responses using elements of DCIs and aspects of CCCs and SEPs (Pellegrino, Wilson, Koenig & Beatty, 2014). Accordingly, an integrated three-dimensional assessment task requires a corresponding integrated three-dimensional rubric so that we can make sense of the learning demonstrated in the responses.

The *Next Generation Science Assessment* (*NGSA*) design process provides a pathway to develop three-dimensional tasks and accompanying rubrics that are supportive of *NGSS* instruction. The process, illustrated in Figure 7.2 and described in the preceding chapters, uses performance expectations (PEs) as the starting point for creating sets of learning performances that guide the development of assessment tasks and rubrics. The process involves six major steps across three phases. The first phase, Steps 1–3, involves selecting a PE or PE bundle and systematically unpacking the dimensions to understand the assessable components. The elaborations from the unpacking are used to create a visual representation in the form of a map, which we refer to as an *integrated dimension map*, that lays out the dimensional terrain for fully achieving the PE or bundle. An integrated dimension map describes key relationships among the DCI elements and identifies how aspects of the SEPs and CCCs can work with these disciplinary relationships to promote students' integrated proficiency (for an overview, see Chapter 3).

In the second phase, Step 4, the integrated dimension map serves as the basis for crafting a set of learning performances that describe the proficiencies that students will need to demonstrate over time as they progress toward achieving a PE or PE bundle. Learning performances are always written as three-dimensional performance statements that integrate DCI elements with aspects of the SEPs and CCCs. Fundamentally, a single learning performance will be smaller in scope than a PE and cover a designated area of an integrated dimension map, thus representing only a portion of the more comprehensive PE or PE bundle (see Chapter 4 for a refresher on how to write learning performances).

When writing a learning performance, we also describe the integrated proficiencies that are required to demonstrate the learning performance and for each we identify the *observable* evidence a student needs to provide to show proficiency with the learning performance. We specify this evidence in a format that we call an *evidence statement*. Learning

performances and their associated evidence statements serve as a keystone in the process of developing tasks and rubrics that enable us to assess "building toward" the performance expectations. A learning performance describes the performance that students need to demonstrate; evidence statements describe what must be evident in students' demonstration of the learning performance. Without evidence statements, we would not be able to determine whether students have met a learning performance (see Chapter 4 for guidance on how to specify evidence statements for learning performances).

FIGURE 7.2. The six steps of the *Next Generation Science Assessment* design process

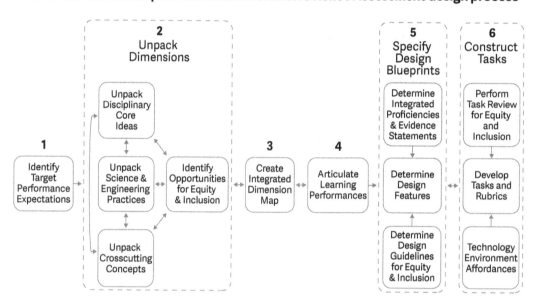

In the third phase, Steps 5–6, we use an organizational strategy called a *design blueprint* to guide the principled development of tasks and rubrics. Design blueprints are documents that serve as all-important organizers for using learning performances to construct assessment tasks to assess for three-dimensional learning (see Chapter 5 for a run-through of design blueprints and direction on how to use them to create tasks). They include five key elements for designers: integrated proficiencies, evidence statements, essential task features, variable task features, and equity and inclusion considerations. *Integrated proficiencies* and *evidence statements* provide designers with a clear picture of what inferences regarding student proficiency are to be made and inform designers of the evidence that is required for a student to demonstrate proficiency with a learning performance. *Essential task features* specify the common features that all tasks must include. *Variable task features* specify the important features that can vary in some form or by option across tasks. *Equity and inclusion considerations* are described so that designers have targeted guidance for developing tasks that are accessible and fair for a wide range of students. The considerations for equity and inclusion are drawn from the unpacking step in the first design phase (refer to Chapter 6 for a deep dive into how equity and inclusion considerations are made and then woven throughout the *NGSA* design process).

Attention to rubrics in the *NGSA* design process begins with the learning performances and their evidence statements in the second phase. Clearly specified evidence statements guide the writing of the tasks and rubrics that will be used to assess whether students have met a learning performance. Design blueprints provide the specification for both tasks and rubrics in the third phase of the process. Noteworthy is that the development of a rubric draws from much of the same source material used to develop the three-dimensional task.

Rethinking Rubrics for Three-Dimensional Learning

Assessment best suited for *NGSS* classrooms will be aligned with three-dimensional learning goals and responsive to instruction (Pellegrino et al., 2014). A well-designed assessment task will create the context for students to demonstrate their three-dimensional learning and a well-designed rubric will provide a clear picture of the various ways that students might show proficiency in their responses. In short, a rubric is an evaluation tool that can be used to make judgments about student performance and progress. Beneficially, rubrics enable teachers to make sense of student responses so a determination can be made about how three-dimensional learning is progressing.

A three-dimensional rubric is a clearly written description, typically organized into a table format, of the performance criteria by which student responses are judged. It contains the *observable* criteria that serves to guide what to look for in student performance. Rubrics that contain criteria for what constitutes performance are ideal partners for three-dimensional tasks because they can be used to describe levels of performance along a scale of high (e.g., proficient or advanced) to low (e.g., beginning or emerging). By including both criteria and a scale in a rubric, students can be assessed for a range of possible performances. Importantly, the information provided by applying such a rubric can be used for monitoring class progress, providing students with feedback, and guiding next steps for instruction. This is different from how rubrics are mostly used in science classrooms, which is to assign scores and grades for accountability purposes. Three-dimensional rubrics can be developed and used for grading too, but the real value of them is that they can be used to improve ongoing classroom teaching and learning.

An important distinguishing feature of three-dimensional rubrics can be found in how they are developed. A very common approach in classroom-based assessment design is to create the rubric directly from the task. In this approach, the task itself is the main source for identifying the performance criteria and generating written descriptions of the possible levels of performance. Oftentimes, the question prompt within the task is the starting point for determining the criteria for the rubric. One pitfall with this approach is that if the task is misaligned to the learning performance in any way or if the question prompt within the task is off the mark, then the rubric will also have this shortcoming. As a result, you may end up with a nicely coupled task and rubric that are aligned to one another, but they might not be appropriately aligned to the learning performance. Most concerning in this scenario is that you would be making judgments of student performance based on the task rather than the learning performance that the task is intended to draw out.

To ensure that rubrics focus attention on what matters for three-dimensional learning, it is important that their development begins with the learning performances. First and foremost, the task and rubric should be developed from the same design blueprint so that they are each aligned to the same learning performance. In this approach, the task and rubric both emerge from what you are looking for in student performance as described by the learning performance. Using the same design blueprint to develop both enables you to have a task and accompanying rubric focused on the same integrated proficiencies. An added benefit is that you can cross-check them against each other and remedy any design issues. For instance, when the source of the rubric is the learning performance rather than solely the task, you can use the rubric to catch any issues with the task itself and vice versa. Importantly, this provides you with another means to confirm consistency among the target learning performance, task, and rubric.

Using a Design Blueprint as the Starting Point to Create a Three-Dimensional Rubric

Clearly specified design blueprints serve as the ground plans for designers to create three-dimensional tasks and accompanying three-dimensional rubrics. At the onset, it is important to read carefully through the key elements of the blueprint—whether the aim is to create a task or rubric—to have a clear understanding of each element and how they are intended to work together to define the coverage or scope of the intended task. When creating a task, designers will draw from all the key elements in a blueprint: integrated proficiencies, evidence statements, essential task features, variable task features, and equity and inclusion considerations. Table 7.1 shows the design blueprint used to create the *Rosy's Battery Under Water* task that was introduced in the scenario at the beginning of the chapter. This design blueprint was made for learning performance LP5 that is part of a set of learning performances aligned with the bundled *NGSS* performance expectations MS-PS1-2 and MS-PS1-5. When creating a rubric, designers will initially focus attention on the integrated proficiencies, evidence statements, and essential task features that are specified in the blueprint. These three blueprint elements are all taken into account to produce a *rubric template* that displays the observable criteria that will serve to guide what to look for in student performance. Alongside the criteria, the rubric template includes a description of the anticipated proficiency levels for each along a scale, typically from proficient to beginning levels. Noteworthy is that the rubric template originates from the design blueprint and serves as the basis for creating a task-specific three-dimensional rubric. A *task-specific rubric* identifies the specialized performance criteria for a particular task. Just as a design blueprint can be used to create one or more tasks, so too can a rubric template be used to create an accompanying three-dimensional rubric for each task created from the same design blueprint.

TABLE 7.1. Task Design Blueprint for Learning Performance LP5 Showcasing the Specifications for Each of the Five Elements Used to Create Rosy's Battery Under Water Task

Learning Performance	LP5: Students develop a model of a chemical reaction that explains new substances are formed by the regrouping of atoms, and that mass is conserved.
Integrated Proficiencies	• Ability to construct a model of a chemical reaction that shows regrouping and conservation of mass. • Ability to describe how a model shows that atoms regroup during a chemical reaction. • Ability to support a model by explaining that chemical reactions conserve atoms and therefore conserve mass.
Evidence Statements	• Students' model shows that atoms are correctly regrouped from reactants to products and each type and number of atoms is conserved. • Students' description of a model show that atoms are rearranged during a chemical process. • Students' statement about a model show that mass is conserved by providing evidence of number of each type of atom before and after a reaction.

(Continued)

Table 7.1. (continued)

Learning Performance	LP5: Students develop a model of a chemical reaction that explains new substances are formed by the regrouping of atoms, and that mass is conserved.
Essential Task Features	• Task presents a scenario involving a simple chemical reaction. • Task provides information about the composition and structure of the substances involved in the chemical reaction. • Task scenario provides information at the atomic levels about the reactant(s) and product(s) in a given chemical reaction. • Task prompts students to construct a model of a chemical reaction. • Task includes a key for students to use to construct a model. • Task provides a way for students to express their model of what is happening at the atomic level during a chemical reaction. • Task provides a prompt to describe how a model shows that atoms rearrange in a chemical reaction. • Task provides a prompt to make a statement about how a model explains why mass is conserved during a reaction.
Variable Task Features	• Task scenarios can vary by phenomena and/or types of chemical reactions. • Task scenarios can vary in how information is presented (e.g., text only, text + picture, visual aids, etc.). • Tasks can require different ways to express the model, such as by drawing, writing, and/or using a modeling software tool. • Tasks can vary in the type and number of atoms to be regrouped from reactants to products in the model. • Tasks can vary in their level of scaffolding for the practice of developing and using models. • Tasks can vary in terms of increased or reduced demands for prior knowledge of chemical formulas and chemical reaction equations.
Equity and Inclusion Considerations	• Select chemical reaction phenomena that are relevant to students' lives and/or interests. • Write a reasonably compelling scenario with a familiar, relevant, or authentic situation that encourages students to engage with the phenomenon and to work through the task. • Use straightforward everyday language when describing information that is not being assessed.

Guidance for Developing Rubric Templates

A rubric template displays the observable criteria or "look fors" of student performance and describes the proficiency levels of performance. The benefit of a template is that it establishes a general framework for evaluating students' proficiency with a learning performance (Harris, McNeill, Lizotte, Marx & Krajcik, 2006). We recommend that you develop the rubric template directly from the design blueprint and do so before you begin to create tasks. By developing the template from the design blueprint, you set an objective basis for judging performance that is strongly aligned to the learning performance. Once a rubric template is developed, you now have the evaluative criteria at hand so that you can readily create the task-specific rubric for each task that is produced.

When developing a rubric template from a design blueprint, we always begin with the integrated proficiencies and evidence statements. Integrated proficiencies describe the abilities that students are expected to employ when doing a learning performance. They provide information about what it is that students should know and be able to do to demonstrate a learning performance. Evidence statements specify the observable evidence to look for in a student's performance. They provide a clear picture of what "counts" as important when students use and apply knowledge to demonstrate a learning performance. We also review the essential task features to pinpoint the requirements for the question prompts that are intended to elicit student performance. Essential task features refer to the attributes that must be included in all tasks created for a particular learning performance. We review this element with an eye on identifying the features that are intended to prompt students to demonstrate the learning performance. Table 7.2 presents the key elements that were drawn from the design blueprint for learning performance LP5 and used in the development of the rubric template.

TABLE 7.2. Three Key Elements From the Design Blueprint for Learning Performance LP5 That Inform the Development of the Rubric Template

Learning Performance	LP5: Students develop a model of a chemical reaction that explains new substances are formed by the regrouping of atoms, and that mass is conserved.
Integrated proficiencies (required to demonstrate the learning performance)	Assessment targets: • Ability to construct a model of a chemical reaction that shows regrouping and conservation of mass. • Ability to describe how a model shows that atoms regroup during a chemical reaction. • Ability to support a model by explaining that chemical reactions conserve atoms and therefore conserve mass.
Evidence Statements (associated with each integrated proficiency)	Response requirements: • Students' model shows that atoms are correctly regrouped from reactants to products and each type and number of atoms is conserved. • Students' description of a model shows that atoms are rearranged during a chemical process. • Students' statements about a model show that mass is conserved by providing evidence of number of each type of atom before and after a reaction

(Continued)

Table 7.2. *(continued)*

Learning Performance	LP5: Students develop a model of a chemical reaction that explains new substances are formed by the regrouping of atoms, and that mass is conserved.
Essential Task Features (relevant for eliciting each integrated proficiency)	Each task must include: • A prompt to construct a model of a chemical reaction. • A prompt to describe how a model shows that atoms rearrange in a chemical reaction. • A prompt to make a statement about how a model explains why mass is conserved during a reaction.

We have found that the most efficient way to organize and present a rubric template is in a table format. The table should display the observable criteria for student performance and describe the proficiency levels along a scale. As a rule, the observable criteria should draw directly from the integrated proficiencies. The scale should provide a written description of the different levels of quality for student performance. We turn to the evidence statements to determine how a student at a proficient level should perform for each observable criterion. Once we describe what a proficient level should look like for each observable criterion, we then sketch out the range of possible performances that extend from *proficient* to *beginning* levels. We recommend that you use no more than three to four levels for your scale. In Table 7.3, we show the rubric template that was created from the design blueprint for learning performance LP5.

We follow two general steps to create the template. To begin, the integrated proficiencies that will serve as the observable criteria for student performance are placed in the first column, with each row earmarked for a criterion. The rubric template in Table 7.3 presents three observable criteria for the learning performance. Then, to the right, the proficiency levels are described, beginning on the far right with a description of how a student at the proficient level should perform. The proficient level description is drawn directly from the evidence statements that correspond to each integrated proficiency. The level *proficient* is typically the highest level of a three-dimensional rubric, though there might be a circumstance in which a more expert or *advanced* level is preferred. Either way, we always lead off by determining how a student who would be considered proficient would demonstrate the learning performance. Once this is accomplished, we complete the remaining columns for each level by describing their distinguishing indicators. The rubric template in Table 7.3 shows three proficiency levels from left to right: *beginning, developing, and proficient.*

TABLE 7.3. Rubric Template Created from the Design Blueprint for Learning Performance LP5

Observable Criteria for Performance	Proficiency Level		
	Beginning	Developing	Proficient
A. Construct a model of a chemical reaction that shows atoms are regrouped from reactants to products and that mass is conserved.	Student model does not show the correct reactant and product molecules; might show or not that atoms are conserved in the process.	Student model shows only the correct reactant and product molecules in a chemical reaction; does not show that atoms are conserved in the process.	Student model shows the correct reactant and product molecules in a chemical reaction and that atoms are conserved in the chemical reaction process.
B. Describe how a model shows that atoms regroup during a chemical reaction.	Student does not indicate how the model shows that molecules in the products are made from the rearrangement of atoms of the reactants.	Student partially indicates how the model shows that molecules in the products are made from the rearrangement of atoms of the reactants.	Student indicates how the model shows that molecules in the products are made from the rearrangement of atoms of the reactants.
C. Support a model by explaining that chemical reactions conserve atoms and therefore conserve mass.	Student does not describe that the model shows mass is conserved; does not support the model with reasoning that atoms are conserved.	Student partially describes that the model shows mass is conserved; might only state that mass is conserved or that atoms are conserved; does not support the model by stating that mass is conserved because the number of all types of atoms remain the same in the process.	Student describes that the model shows mass is conserved by stating that atoms are conserved because both sides of the reaction consist of the same number of all types of atoms.

Using a Rubric Template to Create a Task-Specific Three-Dimensional Rubric

After a design blueprint has been used to produce a rubric template and a task, we then use the rubric template and the task in tandem to develop a three-dimensional rubric that will accompany the task. Whereas the rubric template provides a general framework, the task-specific rubric identifies the specialized performance criteria for the task. Importantly, it details the unique response requirements of the task and specifies the observable differences between proficiency levels.

The rubric template serves as the basis for creating the task-specific three-dimensional rubric. To create the rubric, we use a table format and follow a series of steps that entail using the template alongside the task. First, the same set of performance criteria from the template are placed in the left column and the general proficiency levels of the learning performance, also drawn from the template, are placed in the middle and right columns. Then, the response requirements and proficiency levels of the task are incorporated into the rubric. There are several ways to accomplish this. A straightforward approach is to begin with the proficient level column and for each performance criteria succinctly describe the specific observable indicators of proficiency for the task. In this approach, the indicators are written immediately below each of the general proficiency levels. Once this is done for the proficient level, then the indicators for the remaining different levels of proficiency (e.g., *developing* and *beginning*) are written down. The indicators should describe what must be present or missing for each level. By including the task indicators below each proficiency level, you can easily reference back and forth between the proficiency levels of the learning performance and the specific indicators for what constitutes a full or partial response to the task. Table 7.4 shows a task-specific three-dimensional rubric that was developed using this approach for the *Rosy's Battery Under Water* task. Notice that this task-specific rubric targets only the key task response requirements that align to the evaluative criteria in the template. Depending on your own goals for student performance, you might include other task-specific indicators. For instance, the task prompt asks students to use a key to create their models and you could include how well students use a key as one of the indicators for the "developing a model" performance in the rubric. At minimum, the rubric should bring together all the evaluative criteria described in the template with the corresponding task-specific indicators. Beyond minimum, keep in mind that you can include other task-specific indicators that match with your own goals for student performance.

TABLE 7.4. Three-Dimensional Rubric Expressly Created for Rosy's Battery Under Water

Rubric: *Rosy's Battery Under Water*			
Performance Criteria	**Beginning**	**Developing**	**Proficient**
Developing a Model: Construct a model of a chemical reaction that shows atoms are regrouped from reactants to products and that mass is conserved.	Student model does not show the correct reactant and product molecules; might show or not that atoms are conserved.	Student model shows only the correct reactant and product molecules in a chemical reaction; does not show that atoms are conserved in the process.	Student model shows the correct reactant and product molecules in a chemical reaction and that atoms are conserved in the process.
	Model includes **NONE** of the indicators or **ONE** of the indicators listed under 'Proficient': • Equal number of all types of atoms on each side	Model includes **ONE** of the indicators listed under 'Proficient': • H_2O as the reactant and H_2 and O_2 as the products	Model includes **ALL** of the following indicators: • H_2O as the reactant and H_2 and O_2 as the products • Equal number of all types of atoms on each side
Describing a Model: Describe how a model shows that atoms regroup during a chemical reaction.	Student does not indicate how the model shows that molecules in the products are made from the rearrangement of atoms of the reactants.	Student partially indicates how the model shows that molecules in the products are made from the rearrangement of atoms of the reactants.	Student indicates how the model shows that molecules in the products are made from the rearrangement of atoms of the reactants.
	Response includes **NONE** of the indicators listed under 'Proficient' or 'Developing'.	Response includes **SOME** portion of the indicators listed under 'Proficient': • Atoms regroup/rearrange OR atoms form products (but not both) • Molecules break apart OR new substances are formed (but not both)	Response includes EITHER or BOTH of the following indicators: • Atoms regroup/rearrange to form products • Reactant molecules break apart and form new substances
Supporting a Model Support a model by explaining that chemical reactions conserve atoms and therefore conserve mass.	Student does not describe that the model shows mass is conserved; does not support the model with reasoning that atoms are conserved.	Student partially describes that the model shows mass is conserved; might only state that mass is conserved or that atoms are conserved; does not support the model by stating that mass is conserved because the number of all types of atoms remain the same in the process.	Student describes that the model shows mass is conserved by stating that atoms are conserved because both sides of the reaction consist of the same number of all types of atoms.
	Response includes **NONE** of the indicators listed under 'Developing' and 'Proficient'; may include incorrect information about conservation.	Response includes **SOME** portion of the indicators listed under 'Proficient' and may include **ANY** of the following: • Mass is conserved but does not provide reasoning that atoms are conserved • Atoms are conserved but does not state that the number of O and H atoms remain the same	Response includes ALL the following indicators: • Mass is conserved because atoms remain the same in the process • Both sides of the reaction consist of an equal number of atoms (O and H)

Another approach for organizing a rubric into a useful structure is to merge the response requirements and proficiency levels of the task with the performance criteria and general proficiency levels of the rubric template. In this integrated approach, the general descriptions from the template are rewritten to include the task-specific details. This includes adding details to the performance criteria so that they also specify the criteria of the task. Once the performance criteria are detailed, the general proficiency levels are made more task-specific and include the specific observable indicators of proficiency for the task. When finished, the result is a rubric with clarity and detail for evaluating the range of student performances on the task as defined by the learning performance.

Guidance for Finalizing the Task-Specific Rubric

A great advantage of a task-specific rubric is that it presents a useful structure for evaluating student performance across a range of proficiency levels. This is the most supportive type of rubric for instruction because it can provide descriptive details on what three-dimensional performance looks like. Yet, it is important to keep in mind that determining the various levels to include in a rubric and writing the descriptive details for each will take some time—and likely more than one round of revision. Finalizing a rubric is very much an iterative process where there is a back-and-forth required to achieve a strong match between the learning performance, task, and rubric.

After creating the first draft of a task-specific rubric, we recommend that you pause and review the alignment of the rubric to the design blueprint and to the task and make adjustments as needed. This cross-check is a means to confirm consistency between the task and rubric and to check that they adhere to the important blueprint requirements. First and foremost, the task should elicit evidence of the learning performance, the rubric should provide a structure for evaluating that evidence, and both should be anchored to the specifications in the design blueprint. In doing this review you might find, for example, that some descriptions and indicators in the rubric may need to be revised, or that a prompt within the task may need to be refined so that it more adeptly elicits the integrated proficiencies of the learning performance.

Another valuable next step is to affirm the rubric by collecting samples of student responses and sorting them according to the described proficiency levels. There are several benefits to doing this. First, examining responses that students have completed for a task can provide you with a range of examples that can be used to refine the proficiency levels in the rubric. The range of examples might, for example, serve to affirm the proficiency levels and their descriptions, or they might call attention to some clarifications and revisions that are needed. Second, the responses can provide you with fodder for adding in descriptive detail to the various levels. For instance, from the sample you might identify additional indicators for the levels or additional boundaries such as what is missing in a response that would relegate it to a lower level. Third, typical responses can be selected and used as exemplars that illustrate the various levels on the scale. Adding exemplars to the rubric can be especially beneficial for ensuring consistency in situations where there are large numbers of responses to be evaluated and in situations where there are two or more teachers using the same rubric.

A closing consideration when finalizing a three-dimensional rubric for instructional use is that there can be value in creating a version for students. When students know the performance criteria on which they were judged, they can get a clearer picture of what was expected and take steps to improve. A student-appropriate version is one that uses vocabulary, phrasing, and sentence structures that will be accessible and clear to all students who will be using it. Also of high value are samples of student work that illustrate the proficiency levels. Thoughtfully reconfigured rubrics for students can help make teachers' expectations clear and support three-dimensional learning by showing students what it looks like to use and apply knowledge.

Evaluating Student Responses With a Rubric

If a rubric is to be used instructionally, then its fundamental purpose is to help you make sense of students' multidimensional performance. It should facilitate your understanding of student performance in relation to the integrated proficiencies of the learning performance and enable you to judge varying levels of performance. More broadly—yet just as important—it should contribute meaningful information about how students are making progress in building toward the overarching PE or PE bundle. Returning to the rubric in Table 7.4 that accompanies *Rosy's Battery Under Water*, this rubric can help pinpoint strengths as well as areas for further growth for modeling atom rearrangement and conservation of mass in the context of a chemical reaction. You will recall that the task and rubric were developed for learning performance LP5: *develop a model of a chemical reaction that explains new substances are formed by the regrouping of atoms, and that mass is conserved.* This learning performance covers a major portion of the PE bundle and for this reason the accompanying task and rubric stand to enable a teacher to gain valuable insight into how students are building toward proficiency with the bundle.

The rubric presented in Table 7.4 provides a useful structure for evaluating student performance on developing a model and describing and supporting its explanatory power.

Beneficially, the rubric integrates DCI elements related to chemical reactions and aspects of the CCC of Energy and Matter with the three target aspects of modeling to help a teacher interpret and evaluate students' three-dimensional performance. Figure 7.3 presents a middle-grade student's response to *Rosy's Battery Under Water*. Based on the rubric, how would you rate this response? The student's model shows how the atoms from the water molecules rearranged to form gas molecules. The model also shows that the atoms are conserved in the chemical process. The student's description of the model accounts for the role of electricity in "breaking apart" the water molecules into hydrogen and oxygen gas and includes reasoning for why mass is conserved in the process. This is a sophisticated response for a middle school student and is representative of a response that would be rated proficient for all three performance criteria.

FIGURE 7.3. A student's response that demonstrates proficiency across all three performance criteria

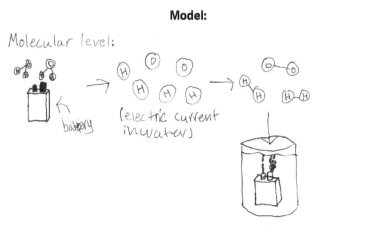

Model:

Molecular level:

battery

(electric current in water)

Description and support of the Model:

explination:
The battery sends an electric current through the water, breaking apart the molecules (H_2O). The attoms rearange into oxygen gas (O_2) and hydrogyen gas (H_2). In this reaction there is the same amount and type of attoms before and after the reaction, so mass is conserved.

"The battery sends an electric current through the water, breaking apart the molecules (H_2O). The atoms[sp] rearrange[sp] into oxygen gas (O_2) and hydrogen gas (H_2). In this reaction there is the same amount and types of atoms[sp] before and after the reaction, so mass is conserved."

Figure 7.4 presents another middle-grade student's response to *Rosy's Battery Under Water*. In this response, the model shows the required elements much like the prior student's model, and appropriately sequences the process from reactants to products. Also similar to the prior student's response, this student describes how the model shows the role of electricity in splitting apart the water molecules and describes that the atoms rearrange to form gas molecules. Notice that differently, this student's response does not include reasoning for conservation of matter. Even though the evidence in the model suggests that the student *might* have applied this reasoning when drawing it, there is no clear or conclusive evidence that the student did. This is an important missing piece worth noting by a teacher because the notion of conservation of mass is a difficult concept for many middle school students to grasp, yet vital for making sense of phenomena across the disciplines and for future learning in science. Despite this missing piece, this is a very strong response for a middle school student and would be rated proficient for two of the performance criteria and at a beginning proficiency level for the third criterion.

FIGURE 7.4. A student's response that demonstrates varying levels of proficiency across the performance criteria

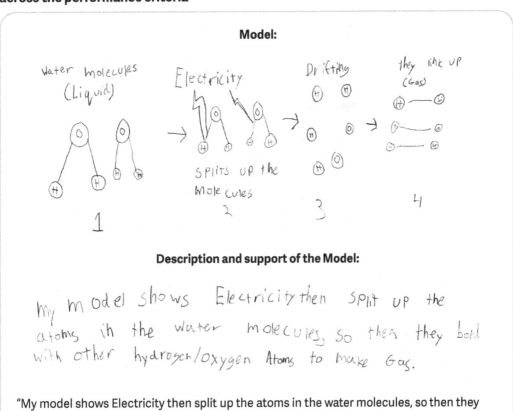

"My model shows Electricity then split up the atoms in the water molecules, so then they bond with other hydrogen/oxygen atoms to make gas."

Figure 7.5 presents a third example of a middle-grade student's response to *Rosy's Battery Under Water*. This student's model shows a water molecule with an arrow pointing to the three atoms that have been separated from the molecule. Also, the model shows the same number and types of atoms on each side. How would you rate this model? The student has appropriately included the reactant and illustrated that mass is conserved, but the atoms on the right side are not rearranged into new molecules. For this reason, the model does not fully serve to visualize and explain the phenomenon. Accordingly, the model would be rated at a beginning proficiency level. The student's description includes an explanation of how the two gases could have come from the water. Notice that this description includes that the water molecule's atoms break apart from each other and become gas, but it does not address that the atoms regrouped. Notably, the description includes some of the

FIGURE 7.5. Another example of a middle-grade student's response to Rosy's Battery Under Water

Model:

Description and support of the Model:

"The two gases could've come from water due to the battery being dropped in. Water has hydrogen and oxygen atoms, and those atoms broke apart and turned into gas due to the battery. This model shows that although the molecule's atom[s] broke apart, there were still the same number of atoms and same type. This means that there is mass conserved in this reaction. Since the number of atoms and type of atoms stayed the same."

appropriate indicators and would be rated at a developing proficiency level for the second performance criterion. Regarding the third performance criterion, the student's description includes reasoning for why mass is conserved in the process and would be rated proficient. Noteworthy is that this student was rated at a different proficiency level for each criterion—beginning, developing, and proficient—thereby pinpointing one strength and two areas for feedback and further growth.

When evaluating many student responses to a task, it can be helpful to create a table to use alongside the rubric to jot down notes about what you are noticing about students' performance. As students' responses are reviewed, the teacher notes individual performance as well trends in performance across students. This organizational strategy can enable the teacher to see similarities and differences in students' performance more easily during the evaluation process. Once all the responses have been reviewed, the result is a summary table that can be used to identify patterns that may indicate areas where students as a class are doing well and other areas where performance could be improved. Beneficially, this information can be used as the basis for providing feedback to students. Moreover, this information can be helpful for planning next steps for classroom instruction.

Advantages of Integrated Multidimensional Rubrics

A rubric structure that calls attention to the specialized performance criteria for a task has several advantages over a holistic rubric structure that results in a single score for a multidimensional student response. First, this structure makes it efficient for a teacher to interpret and evaluate students' three-dimensional performances by making the performance criteria explicit. Second, this structure enables a teacher to focus on one performance criterion at a time when evaluating a response, without the need to consider multiple aspects of performance all at once. This makes it easier to apply the rubric to a complex multidimensional response and can make the evaluation process more manageable, consistent, and reliable. Third, because each performance criterion is aligned to a specific integrated proficiency, teachers can more accurately determine what students know and can do and more readily use this information to support learning.

The rubric for *Rosy's Battery Under Water* task is a relatively straightforward illustration of how multiple rubric components can explicitly connect with key aspects of the design process such as essential task features, integrated proficiencies, and evidence statements. Also worth noting is that rubric components can be helpful for disentangling distinct proficiencies that could more easily be conflated within a task. For example, an

assessment task might prompt students to construct an explanation about the *how* or *why* of a phenomenon. In responding fully to this prompt, students may be expected to use important aspects of the practice of constructing explanations that includes supporting a claim with both evidence (i.e., from data) and reasoning (i.e., based on underlying science principles). Denoting separate rubric components for the important aspects can bring forward the extent to which an individual student (or the collective class) is able to use evidence with reasoning in support of a claim and help to pinpoint aspects where further instructional support may be needed.

Finally, using this rubric structure focuses the assessment process on the multiple *NGSS* dimensions that reflect the integrated, three-dimensional nature of the learning performance (and associated larger PE). This kind of rubric does not separate the dimensions from one another—all judgments of a student's performance reflect the extent to which they have *integrated* the SEP, CCC, and DCI dimensional aspects and elements in their response. A consequence of this design is that students' performance on a three-dimensional task as reflected in a multidimensional rubric may not appear as high relative to performance on a more typical content-focused (i.e., one-dimensional) task. This is because demonstrating three-dimensional learning is more complex and thus demonstrating knowledge-in-use raises the bar for student performance. With an integrated rubric, students *must* exhibit some multidimensional proficiency to achieve a developing or proficient rating on a rubric component; students cannot achieve a proficient level with a one-dimensional response. For this reason, it is important to keep in mind that the real value in creating and using this kind of rubric is to gain insight into how well students can integrate the multiple dimensions of science knowledge as they make progress toward achieving the performance expectations.

Using Three-Dimensional Rubrics in Instructionally Supportive Ways

Once students have responded to a task, how can their responses best be interpreted to inform instruction? An advantage of a well-designed rubric is that it establishes an objective basis for making judgments about student performance (Arter & McTighe, 2001). It can bring consistency and efficiency to the work of reviewing the responses of many students. Moreover, when reviewing many responses, a rubric can help bring to light patterns in students' responses that may indicate not only what they have learned thus far, but also what instruction might be needed next. Yet, all too often, rubrics are used primarily for grading purposes (Abell & Siegel, 2011). When used mostly for grading, teachers miss out on the most powerful benefits to using rubrics. For example, teachers can use rubrics to gain greater insight into students' range of performance and deepen their own professional knowledge of three-dimensional learning. Knowing what three-dimensional looks like and the various ways that students can demonstrate proficiency can, in turn, inform classroom instruction. This example represents one of six strategic uses of rubrics shown in Table 7.5 that extend beyond grading and toward using three-dimensional rubrics in instructionally supportive ways. In this section, we describe these six strategies and how they can be used to inform teaching practice and promote student learning.

TABLE 7.5. Six Strategic Uses for Rubrics Beyond Grading

- **Diagnose** individual and collective students' proficiency, keeping an open mindset for unconventional responses that express important scientific ideas and practices, to provide targeted feedback that advances learning
- **Follow** student progress over time to inform ongoing instruction
- **Spark** productive class discussion to engage all students in building their science knowledge with one another
- **Promote** student self-assessment and/or peer-critique to activate student agency for learning
- **Encourage** reflection on one's own teaching practice to better meet student needs
- **Open** a window on student performance to deepen professional knowledge of three-dimensional learning

Diagnose Students' Proficiency

Among the most beneficial uses of rubrics is to diagnose individual and collective students' proficiency so that targeted feedback can be provided to advance learning (Arter & McTighe, 2001). When used to focus in on individual learning, a rubric can draw attention to integrated knowledge-in-use and help pinpoint what a student does well and what may need improvement. To accomplish this in a manageable way, it is important to use the rubric to focus on the high-value aspects of performance so that you can identify what could be improved and decide on the feedback that is needed. By *high-value*, we mean the aspects of performance that will have greatest impact for learning in your classroom. Note that a well-designed rubric can enable you to identify many strengths as well as areas for improvement, but you will need to decide on what will be most beneficial for students. Important to keep in mind is that students may express their ideas and performances in diverse ways and in some of these instances the responses may seem less conventional.

When applying your rubric, it is critical to maintain an open mindset, accepting unconventional responses to see if students express important scientific ideas and practices.

When used to focus in on the collective learning of a class, a rubric applied across many responses can call attention to patterns in students' performance. These patterns may indicate areas where students as a class are doing well and indicate other areas where performance could be improved. Patterns indicating both types of areas can be helpful for planning classroom instruction more effectively. For example, a rubric might confirm that students are doing well with developing models but bring to light an area of challenge in using models to explain the *how* or *why* of phenomena. A next-step instructional decision for the teacher could be to place students into small groups to develop a group model of a phenomenon and then use it to present to the larger class an explanation for a how or why question about the phenomenon. The teacher could invite the class to ask questions about each model and what it is intended to explain about the phenomenon, provide targeted feedback to the student presenters, and give each group the opportunity to revise their models and explanations.

A major benefit of using a rubric to look at both individual and collective students' proficiency is that it can be the basis for providing more tailored and specified feedback that can guide individual students or the whole class toward revisiting aspects of ideas or performance that are off the mark or to attending to aspects that have been identified as missing. Notably, the best type of feedback will be actionable and given in a student-friendly manner. For feedback to be most beneficial, it needs to be actionable so that students know how to improve. Finally, it is important to remember that feedback can also be beneficial when it identifies a strength or success. When students do well, such as successfully demonstrating a complex learning performance or showing progress and improvement, this type of information can make students more aware of what they are learning and better equipped to build on their success.

Follow Student Progress Over Time

A benefit of using tasks that are designed from learning performances is that you can place them in your instructional sequence at various time points where it will be advantageous to check in on student progress. With a set of tasks placed across a sequence of instruction, a teacher can use rubrics to follow students' progress over time as they build proficiency toward a performance expectation or bundle of performance expectations. In turn, this information can be used to inform ongoing instruction. To arrange for this to happen requires that formative assessment be an integral part of the instructional planning process. This entails selecting tasks that align well with instruction and then planning for when and how you will embed them into the instructional sequence.

A recommended first step is to begin with the performance expectation or bundle of performance expectations that are the summative targets for three-dimensional learning in your instructional unit or sequence. Next, select the accompanying learning performances (see Chapter 4) that will serve as the intermediary performances that students will need to achieve as they progress through instruction and as they build their proficiencies toward the summative performance expectation or bundle. A third step is to identify the time points in the unit or sequence where each learning performance should be assessed. The final step is to identify a task for each learning performance and place the tasks in your instructional sequence at the appropriate time points. Admittedly, this strategy requires some additional effort in planning, but the gain in doing this is that the teacher is better poised to follow students' progress and catch students who veer off the intended pathway. Importantly, the tasks and rubrics are also sequenced so that the teacher can look for patterns in students' performance over time.

A side benefit of making this a part of your instructional planning process is that it can also nudge you to consider far in advance how you will respond during instruction in those instances when students may not be progressing as expected. Considerations might include teaching strategies or next instructional steps to move learning forward. Another side benefit is that using a sequenced set of rubrics can provide evidence for how students are building proficiency over time that can be used to document their trajectories or pathways for learning. In doing so, this may also deepen your knowledge of the various ways students can build their proficiencies with using and applying the dimensions over time and toward the performance expectations.

Spark Productive Class Discussion

Rubrics, including the evidence and feedback gained from them, can be used to promote productive class discussions that will lead to improved performance and greater understanding about phenomena. Such discussions might take the form of debating the features of what would make for a high-quality response to a task, evaluating two or more competing claims made about a phenomenon in a task, or determining steps to take to improve a response. For instance, a teacher might present an example response (i.e., not an actual student's response) for students to critique and debate the acceptability of a claim, model, or explanation. Or a teacher might use students' ideas from their responses as starting points for a discussion about a phenomenon that will support students in rethinking their expla-

nations. Also beneficial is using patterns found in students' performance to make important aspects of performance more explicit for students. This might entail, for example, a discussion on what criteria makes for asking good investigative questions, constructing high-quality explanations, or developing explanatory models. These types of discussions are of value because they can encourage students to reflect on their own and others' performance, build their science knowledge with one another, and help them to reach consensus with one another and the teacher on what three-dimensional performance looks like.

To benefit from these forms of class discussion, students must be willing to listen and talk, justify their perspectives, and feel comfortable making and receiving critiques for the purpose of improving their thinking, reasoning, and performance. For teachers, it is important to move away from the conventional call-and-respond dialogue in which the main purpose for debriefing students' assessment task performance is for the teacher to present on what was done right or wrong. Instead, the shift is toward asking questions that encourage reflection and drive students to evaluate and improve their performance. In this format, the teacher might restate or revoice a student's contribution to emphasize or clarify ideas, ask students to take a position either as individuals or as a group, or encourage reflection by asking students to listen, compare, and evaluate with an emphasis on collegial discussion and continuous improvement.

Promote Self-Assessment and Peer-Critique

Rubrics can also be strategically used with students to actively involve them in their own learning. Student-appropriate rubric versions, as discussed earlier in the chapter, can enhance students' engagement with learning by directing their attention to information that they can use to improve. For example, when placed into the hands of students, rubrics and the feedback associated with them can support self-reflection and encourage students to rethink and revise artifacts such as their explanations, arguments, and models. Also, when self-assessment is promoted regularly in the classroom, students can begin to consider how and how much they have improved over time. To encourage self-assessment, the teacher might provide students with a question to guide their use of the rubric, such as "What would you like to improve or change?" or "What is a strength or weakness that you see in your response?" or "What is something you are wondering about?"

Rubrics can be useful for peer-critique, in which students give feedback to one another. This can benefit students in several ways. First, students who provide feedback learn about how others approach their work and how they demonstrate three-dimensional performance. This can help students gain greater awareness about how they themselves can take steps to improve. Second, students who receive feedback get a different perspective from a peer whose input is valued. This can help students become more open to rethinking and revising their work. Third, having students use rubrics to provide feedback to one another along with the opportunity to redo or improve upon their work is more like science as a profession where revisions are made based on new evidence and communication with science colleagues. This can help students appreciate the value of collaboration in reviewing evidence, evaluating claims, and improving their own and others' ideas.

Although using rubrics for self-assessment and peer-critique can encourage students to become involved in their own learning, rubrics are not regularly used in this manner in classrooms. For this reason, teachers often find that students need calibrated support for this strategy to be beneficial. Students may need to be supported in becoming open and responsive to feedback, comfortable with evaluating their own performance, and thoughtful and encouraging in their own critiques of classmates' performance. To pave the way for this type of classroom environment, new norms for classroom participation may be needed. One way to support this is to combine the rubric with techniques such as statement starters, that can lead students more directly toward positive and constructive self- and peer-focused evaluation and feedback. The long-term payoff is that students become more reflective, self-regulated, and collaborative learners.

Encourage Reflection on Teaching Practice

The purposeful use of rubrics can encourage teachers to become more reflective in their instructional practice. Developing and using rubrics nudges a teacher to consider learning goals and what counts as evidence for three-dimensional performance. Subsequently, evidence of students' performance from rubrics can be used to reflect on instruction and think about ways to revise or adapt lessons for students. This can lead to insights about how best to move learning forward by adjusting teaching strategies or next instructional steps. For example, using an assessment task with an accompanying rubric after a sequence of activities or lessons can shed light on the ways in which instruction has helped students make progress in building their capacity to use the three dimensions together. Having an idea of how students have come to a particular place in their learning can be invaluable for instructional planning.

In the midst of reflection, teachers become learners and innovators. They spend time examining student responses for patterns and use those patterns to better understand the impact of their instruction. Teachers then use what they learn about individual and collective student progress to tailor ongoing instruction to meet students' specific needs.

A benefit in doing this is that teachers become more adept in taking students' thinking, reasoning, and performance into account when planning. In this process, teachers innovate on their practice and arrive at a deeper understanding of what is needed to actively support three-dimensional learning.

Deepen Professional Knowledge of Three-Dimensional Learning

Another helpful strategic use for rubrics is to make three-dimensional performance visible for teachers so that they, in turn, can deepen their professional knowledge. Although rubrics are widely recognized as valuable tools to evaluate and support student learning, they are also beneficial for building one's own knowledge for what three-dimensional learning looks like. Noteworthy is that by using rubrics for both purposes, you maximize their usefulness as tools to support learning.

By applying well-designed rubrics with clear performance criteria to examine many responses, you stand to increase your awareness of the various ways that students can demonstrate their proficiency. This can be facilitated by being intentional in how you proceed through the review process. For example, an effective guiding question to ask while reviewing students' responses is, "What am I learning about three-dimensional learning?" This straightforward question can encourage you to pause and consider the bigger picture of what is meant by three-dimensional learning and what it can look like for your students to use and apply knowledge. By doing this over time, you iteratively build your professional knowledge of three-dimensional performance and clarify your vision for students' learning. Another guiding question to ask is, "What am I recognizing about the three-dimensional learning of my students?" This question also encourages a "bigger picture" perspective. The added value in taking time to do this is that learning becomes evident in ways that can deepen your understanding and appreciation of students and their progress. In short, you come to know your students better. This can help you to further recognize how students' thinking, reasoning, and performance can be important and valued resources for teaching and learning.

Primary Takeaways

In this chapter, we described how the *NGSA* design process can be used for developing rubrics that chart evidence of how students are progressing in building three-dimensional proficiency. Rather than aim to assess students' proficiency with each dimension separately, the *NGSA* design process goal is to measure students' integrated three-dimensional proficiency. This is of value because it stays true to the *Framework* vision of science proficiency as bringing the three dimensions together in an integrated manner to make sense of phenomena or solve problems (National Research Council, 2012). The chapter also offered practical guidance for how to use rubrics to gain insight into student performance and use the information to go beyond grading and toward improved teaching and learning. All told, there are five main takeaways from this chapter:

1. Learning performances not only help designers to identify key proficiencies and design assessment tasks that will elicit three-dimensional responses, but also inform the construction of rubrics that stand to help teachers interpret individual and collective student responses. With a task designed from a learning performance, a student can demonstrate progress in building toward aspects of a PE. With an accompanying rubric designed from the same learning performance, a teacher can assess that progress.

2. An integrated three-dimensional assessment task requires a corresponding integrated three-dimensional rubric. A well-designed rubric will illustrate what three-dimensional science performance looks like. It will describe criteria for judging student responses and maintain a focus on evidence related to the integrated proficiencies for a learning performance. Practically speaking, a critical role of a rubric is to enable teachers to make sense of student responses so a determination can be made about how three-dimensional learning is progressing.

3. Design blueprints serve as the ground plans for creating three-dimensional rubrics. At the outset, the design blueprint is used to produce a rubric template that establishes a general framework for evaluating students' proficiency with a learning performance. The template displays the observable criteria for student performance and describes the proficiency levels along a scale. It is used as the basis for creating task-specific three-dimensional rubrics. A task-specific rubric identifies the specialized performance criteria for a particular task. Once a rubric template is developed, it can be used to create the three-dimensional rubrics for all the tasks produced from a design blueprint.

4. Rubrics can benefit both teachers and students. For teachers, rubrics can help make accurate and equitable judgments about student performance. Importantly, they can clarify the criteria for making judgments, serve to differentiate varying levels of performance, and help pinpoint individual and collective strengths as well as areas for further growth. For students, rubrics can provide

feedback that will help them learn about their own progress which, in turn, they can use to guide their ongoing learning. Also, when students know the criteria from rubrics, they get a clearer view on what is expected of them and what counts as three-dimensional performance.

5. Rubrics can be used as tools for teaching and for promoting student learning. To accomplish this, teachers need to consider strategic uses for rubrics that will help them meet their learning goals, inform their teaching practices, and improve their instruction. For instance, teachers must be prepared to respond when students are not progressing as expected. Evidence and feedback from rubrics can be used to move learning forward by adjusting teaching strategies or next instructional steps. Rubrics can also be strategically used with students to actively involve them in their own learning. For instance, when placed into the hands of students, rubrics and the feedback associated with them can support self-reflection and encourage students to rethink and revise their work including explanations, arguments, and models. For this to occur, students need to be supported in becoming open and responsive to feedback, comfortable with evaluating their own performance, and thoughtful and encouraging in their own critiques of their classmates' performance. Finally, assessment is a process through which teachers can learn themselves—the thoughtful use of rubrics can encourage reflection on teaching practice and deepen professional knowledge of three-dimensional learning.

References

Abell, S. K. & M. A. Siegel. 2011. Assessment literacy: What science teachers need to know and be able to do. In *The professional knowledge base of science teaching*, ed. D. Corrigan, J. Dillon & R. Gunstone,. Netherlands: Springer Dordrecht.

Arter, J. and J. McTighe. 2001. *Scoring rubrics in the classroom: Using performance criteria for assessing and improving student performance*. Thousand Oaks, CA: Corwin Press.

Harris, C. J. et al. 2006. Usable assessments for teaching science content and inquiry standards. In *Assessment in science: Practical experiences and education research*, ed. M. McMahon, P. Simmons, R. Sommers, D. DeBaets & F. Crowley, 67-88. Arlington, VA: National Science Teachers Association Press.

National Research Council (NRC) 2012. *A framework for K–12 science education: Practices, crosscutting concepts, and core ideas*. Washington, DC: National Academies Press.

NGSS Lead States 2013. *Next Generation Science Standards: For states, by states*. Washington, DC: National Academies Press.

National Research Council (NRC). 2014. *Developing assessments for the Next Generation Science Standards*. Washington, DC: National Academies Press.

CHAPTER 8

Using Assessment Tasks to Promote Student Learning

Consuelo J. Morales, Michigan State University • Sania Zahra Zaidi, University of Illinois Chicago • Joseph Krajcik, Michigan State University

I've Got an Excellent Assessment Task. How Do I Make Use of It?

In the previous chapters, you learned how to develop assessment tasks and rubrics that align with the performance expectations in the *Next Generation Science Standards* (*NGSS*; NGSS Lead States, 2013). But once you have a good task, how can you use it to learn about your student's proficiency in using the three dimensions of science and engineering practices (SEPs), disciplinary core ideas (DCIs), and crosscutting concepts (CCCs); improve your teaching; and support students in pushing forward their learning? This chapter uses three scenarios to explore how you can use assessment tasks during instruction to learn about students' performance to support their three-dimensional learning and what you can do to help students push their knowledge of the three dimensions deeper.

Oftentimes, science teachers give assessment tasks so that they can make an evaluation and provide students with grades. Although grading is a necessary part of teaching and provides some information to students, the teacher, and other key players such as guardians and policymakers, grading is one of the least worthwhile ways to use assessment tasks. More importantly, teachers and students can use assessment tasks to promote learning. Unfortunately, most students in today's science classrooms do not know how to

use the results from assessment tasks to improve their knowledge. The stress on grading puts a focus on students just seeing the results as final and not on helping them improve. To change this situation we as teachers need to change how we think about and use classroom-based assessments with students. Martin Haberman (1995), an expert in education renowned for his research on teachers and teaching, pointed out that exemplary teachers serving diverse students in low-resourced schools spend little time on tests and grading but instead focus more on students' efforts and accomplishments. They listen, observe, and strive to understand their students so that they can better support learning. In science classrooms, this should include skillfully helping students use embedded, formative assessments to advance their proficiency in the three dimensions of scientific knowledge needed to explain phenomena and solve complex problems.

In this chapter, we attend to the question, *How can you and other teachers use assessment tasks during instruction to learn about students' performance and support their three-dimensional learning?* To address this question, we provide suggestions on using assessment tasks to promote student learning of the three dimensions of scientific knowledge. We also examine three primary strategies for using assessment tasks during instruction and the insights they provide about how students are developing in their proficiencies to inform teaching and learning:

1. Follow students' progress as they build proficiency toward three-dimensional learning goals.

2. Check if students can use their knowledge in new situations.

3. Encourage self-reflection and dialogue to promote learning.

Using assessment tasks in these three ways can provide you and your students with meaningful opportunities to learn. Importantly, these uses can offer teachers and students a rich source of information on how students are progressing in building proficiency toward meeting a performance expectation (PE) or PE bundle. Teachers, for example, can support their students in learning by using this information to provide appropriate and actionable feedback. Teachers can also develop their professional knowledge by engaging in the assessment process. When teachers deeply comprehend the PE or bundle, the associated learning performances, and their corresponding evidence statements, a significant benefit is that teachers then have the science background knowledge of the three dimensions to support student learning. For instance, a deep comprehension of the learning goals can help teachers become better attuned to how students can build toward achieving them. Figure 8.1 shows the six steps of the *Next Generation Science Assessment* design process that teachers and designers can use to design tasks that align with the PEs. Before writing the task, the designer needs to deconstruct the performance expectations to develop learning performances and corresponding evidence statements, as these constitute critical steps in designing assessment tasks. These steps reinforce your understanding of the disciplinary core ideas, scientific and engineering practices, and crosscutting concepts in the PEs and the learning performances.

FIGURE 8.1. The six steps of the *Next Generation Science Assessment* design process

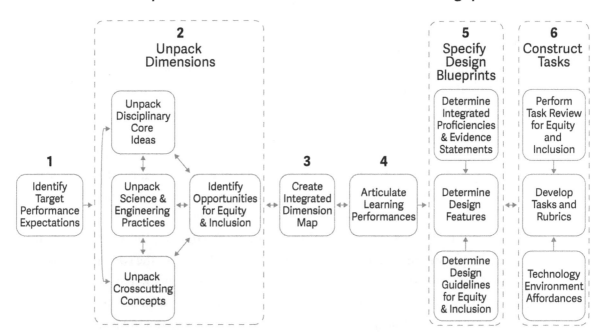

Below we illustrate the uses of assessment tasks using composite fictional scenarios that stem from synthesizing our experiences in observing and working with teachers in middle school classrooms.

Scenario 1: Follow Students' Progress as They Build Proficiency Toward Three-Dimensional Learning Goals

In this scenario, we examine how one 7th-grade science teacher, Ms. McGinnis, used a three-dimensional assessment task to see if her students had developed proficiency in the learning goals she selected. Her class was involved in a unit to build proficiency toward the following two PEs in physical science:

- MS-PS1-2. Analyze and interpret data on the properties of substances before and after the substances interact to determine if a chemical reaction has occurred.
- MS-PS1-5. Develop and use a model to describe how the total number of atoms does not change in a chemical reaction, and thus mass is conserved.

Ms. McGinnis engaged her class in the following driving question: *How do I know if I have some new stuff?* The unit was designed to engage students in exploring if new substances form by investigating the characteristic properties of substances. She began the unit by involving her students in figuring out how they would know if two samples that look similar are the same substance or if they are different. Next, the class explored how substances have unique properties that can identify them as unique substances. Finally, she challenged her students to figure out if two products that looked similar were the same or different substances. To build this proficiency, Ms. McGinnis showed her students two samples of solid white materials and challenged them to figure out if they were the same substance or not. One sample was a nonsmelling soap, and another a piece of paraffin, although Ms. McGinnis did not tell her students the identity of either sample. Her goal was for students to build proficiency toward the learning performance, *Students construct a scientific explanation to explain how they know if two or more samples are the same substance based upon patterns in the properties of substances.*

To build proficiency in this learning performance, students collected melting point data as well as density, odor, and solubility data on each piece of the two unknowns. Once students collected the data, Ms. McGinnis directed them to work in groups to construct a scientific explanation. She reminded the class that scientific explanations always provide a response to a scientific question and need to include a claim that relates the dependent variable to the independent variable(s), sufficient and appropriate evidence to support the claim, and reasoning that justifies why the evidence can support the claim. Students investigated the question of whether the sample materials were made up of the same or different substances. Several students presented their explanations to the students in the class, and they provided feedback to the presenting groups through discussion. Other nonpresenting groups used this feedback to improve their scientific explanations.

Although Ms. McGinnis was pleased with the work students did in groups and during the class discussion, she decided to gather additional evidence to determine if students were building proficiency toward the learning performance. She mused that if students did not know how to identify whether substances are different from each other, they would get confused in determining if a chemical reaction occurred or not in the next segment of the unit.

When developing and planning her unit, she unpacked the performance expectations and created an integrated dimension map to craft a set of learning performances. Next, she used the learning performances to help her sequence the unit. She also identified the time points in the unit where she could use a classroom-based task to see if students developed proficiency of the learning performance. She then designed assessment tasks to provide herself and her students with information about how they were progressing in their integrated learning during the unit. However, while teaching the unit, she recognized that she needed to create another task for her learning performance on distinguishing substances. This led her to craft an evidence statement, identify a related phenomenon, and construct a scenario. As a result, she developed the following evidence statement:

> Students' explanations should include a claim stating that the samples are different substances or the same. They need to include evidence about the properties of the substances that supports the claim as to whether two or more samples are different or the same. Students need to include reasoning by stating that different substances have different properties (i.e., density, melting point, boiling point, odor) that uniquely identify one substance from another.

She reexamined the unpacking document, the learning performance, and the evidence statement in thinking about what task she could develop to provide evidence of student learning. Then, she used the design blueprint to confirm the essential features for the task and consider the variable features she might use. Although it took her several tries, she finally settled on a task in which students would construct an explanation about how they would know whether a collection of metal pieces are the same substance based on their characteristic properties. Figure 8.2 shows the task she developed and named, *Which Metals Are the Same?*

FIGURE 8.2. Ms. McGinnis' physical science assessment task: Which Metals Are the Same?

Ibrahim found four pieces of metal as he was helping his family pull weeds from their garden. He wondered if they were the same or different. His family encouraged him to apply what he had learned in science class.

He first cleaned off the dirt and then observed and recorded their appearance. They all looked similar. He recorded the information in a table. He then measured the mass, volume, and melting point. Using the data, he calculated the density of each piece of metal. He added this information to his data table below.

Ibrahim's Data Table:

Metal Sample	Appearance	Mass	Melting Point	Density
1	Silvery	150g	660°C	2.70 g/cm³
2	Bluish pale gray	110g	692°C	7.14 g/cm3
3	Silvery	150g	962°C	10.5g/cm³
4	silvery	120g	660°C	2.69 g/cm³

Ibrahim then analyzed his data to answer his question, *Are any of the metal samples the same?* Construct a scientific explanation that responds to Ibrahim's question and tells how you know whether the metals are the same or different.

The task requires students to have a high degree of proficiency in using the three interconnected dimensions of scientific knowledge to make sense of phenomena. Ms. McGinnis was pleased with this task because it provided an interesting and realistic scenario and aligned with her instructional goals. Her students would need to apply the following elements from the three dimensions of scientific knowledge to respond to the task:

- Element of the disciplinary core idea: Each pure substance has characteristic physical and chemical properties (for any bulk quantity under given conditions) that can be used to identify it.
- An aspect of the science and engineering practice: An explanation includes a claim that relates how a variable or variables relate to another variable or a set of

variables. A claim is often made in response to a question, and in the process of answering the question, scientists often design investigations to generate data.

- An aspect of the crosscutting concept: Macroscopic patterns are related to the nature of microscopic and atomic-level structures (patterns).

She next developed a task-specific rubric that identifies the observable criteria or "looks fors" of student performance and describes the proficiency levels of performance (see Chapter 7 for an overview of how to develop task-specific rubrics).

The next day in class, Ms. McGinnis gave her students the task and asked them to write an explanation independently. She then instructed the class to pair up with another student to receive feedback from each other, compare their responses, and make modifications to their explanations. She then handed out a student-appropriate version of the rubric to allow students to use the rubric to compare their explanations to the rubric. Finally, Ms. McGinnis had the students reflect on and write out how their explanation related to the rubric.

Ms. McGinnis' instructional decisions have some important elements to them. First, she had students work alone to spark independent thinking. Providing opportunities for students to think through their responses helps them develop integrated knowledge of the three dimensions. Next, she had students read over their explanations to compare them and provide feedback to one another. Working collaboratively with others can support students in building integrated knowledge. When students have opportunities to express their ideas verbally and exchange ideas with others, they build deeper integrated knowledge. Moreover, when students reflect on their responses and compare what they produced to what others constructed, it helps them demonstrate their knowledge of the three dimensions of scientific knowledge and how their ideas need to change.

That afternoon, Ms. McGinnis looked over the explanations. Although she was pleased that students used the various properties as evidence to support their claim that samples one and four are the same substance because they have the same melting and boiling point, she also noticed that some students still showed confusion about whether mass can determine if one sample is the same or different in comparison to other samples. Some students claimed that samples one and three might also be the same because they had similar appearances and the same mass. She thought these students were likely not attending to all the relevant patterns in the data table. Therefore, they might still need support using their knowledge about characteristic properties to distinguish appropriate from irrelevant patterns. This insight nudged her to consider that the class might benefit from a next-step activity to strengthen their thinking and reasoning about characteristic

properties and how to identify one substance from another. Ms. McGinnis decided that the next day in class, she would have students provide explanations for why two samples of similar mass do not provide evidence for whether the two samples are the same substance.

Reflection on Scenario 1: Follow Students' Progress as They Build Proficiency Toward Three-Dimensional Learning Goals

Ms. McGinnis used this assessment task to determine if students were building proficiency toward the learning performance and to support her students in learning some challenging and essential ideas in middle school physical science. She used the task to check in on students' progress by reviewing their responses to identify patterns that indicated what students had learned and determine what instruction she should use next. This is one example of how an assessment task and classroom instruction can work together to provide information to the teacher and the student about their progress in developing proficiency toward a performance expectation, i.e., promote three-dimensional learning. Although the task does not contain all the components of the performance expectation bundle, it does align with the learning performance Ms. McGinnis had constructed, which covers part of the multidimensional terrain of the bundle: *Students construct a scientific explanation to explain how they know if two or more samples are the same substance based upon patterns in the properties of substances.*

The instruction and the task support students in building proficiency toward the PE. In this case, classroom instruction and the task work together synergistically to help learners construct knowledge of the three dimensions of scientific knowledge and to provide feedback to the teacher and student.

Why is it so critical that instruction and assessment go hand in hand? First, even a carefully designed assessment task cannot assess what students *know* and *can do* if students do not have sufficient instructional opportunities to learn. To promote learning, we need to provide opportunities for students to engage with compelling phenomena that require using and applying their knowledge. In this case, Ms. McGinnis was careful in planning instruction that built toward a learning performance, a smaller three-dimensional claim related to the PE, and she also was careful in designing the assessment task. Her instructional decisions provided students with the opportunity to learn, and the task offered valuable information on student knowledge development. In this case, the instruction and the assessment task aligned with each other. Second, without understanding how an assessment task relates to instruction, it is difficult to discern if students are making productive headway toward developing proficiency in a three-dimensional learning goal.

Designing a well-specified assessment task and an accompanying rubric is just half of what is needed to support students' engagement in three-dimensional learning; the other half involves understanding how the assessment task relates to and supports your instruction. Notice how Ms. McGinnis designed her instruction to allow students to reflect on their own and other students' responses and verbalize their integrated knowledge of the three dimensions. Using reflection and dialogue as part of the assessment process helps us move classroom-based assessments away from students seeing them as evaluations and toward seeing and using them to improve their learning. Scenario 3 will further examine the importance of reflection and classroom dialogue in building student knowledge.

This first scenario shows the use of a three-dimensional task that a teacher designed and used to determine if her students were developing proficiency in the three dimensions of scientific knowledge. In addition, it illustrates the importance of classroom instruction aligning with an assessment task to provide the teacher and students with useful and valuable information about three-dimensional learning. Finally, the scenario demonstrates some good instructional techniques by having students work alone, then pair with another student to receive feedback on their work, and finally discuss the task and what makes for a high-quality response in class.

Scenario 2: Check If Students Can Use Their Knowledge in New Situations

The *NGSS* represents a significant shift in the goals of science instruction by putting forth performance expectations that move students away from just knowing *about* science ideas toward *making sense* of science phenomena by using and applying the three dimensions of scientific knowledge. When students can apply their knowledge in new situations, they have developed knowledge they can use when encountering a novel, unexpected, and previously not experienced phenomenon. The *Framework for K–12 Science Education* (NRC, 2012) and the *NGSS* propel us to support students in developing usable knowledge. For learners to develop usable three-dimensional knowledge, they need to use their knowledge in varied situations. This second scenario illustrates an example of an eighth-grade middle school teacher, Mr. Hopper, making use of a task to see if his students can apply their knowledge in a new situation.

Mr. Hopper's students were in the midst of a unit exploring the role of photosynthesis in the cycling of matter and the flow of energy into and out of organisms. Mr. Hopper's unit focused on building proficiency toward the following PE:

• MS-LS1-6. Construct a scientific explanation based on evidence for the role of photosynthesis in the cycling of matter and the flow of energy into and out of organisms.

The unit explores the driving question, *How do plants get the food they need to survive?* In the unit, students collected data showing that plants use the energy from light, carbon dioxide, and water to make sugar (i.e., food) and oxygen. Students performed the following three experiments (See Krajcik and Nordine, 2016, for further details about these investigations):

1. Students gathered evidence that plants use light to produce food and store this food as starch by observing how a germanium plant grows with and without sunlight. The students tested the leaves for starch using an iodine solution which turns a dark purple in the presence of starch. This experiment also provided students with evidence that light is needed for plants to produce starch.

2. The students in the class also wondered what evidence they could collect to determine if plants give us oxygen. To test if plants give off oxygen, the students measured the amount of dissolved oxygen in water using a dissolved oxygen probe. Students placed some Elodea, a plant that grows in water, in a test tube filled with water. Then they measured how the dissolved oxygen changed over time compared to a test tube without Elodea.

3. They also wanted to gather evidence to test if plants used carbon dioxide during photosynthesis. To gather evidence that plants use carbon dioxide, students placed some Elodea in a test tube filled with water but saturated with carbon dioxide. Students used Bromothymol Blue (BTB), a qualitative indicator for carbon dioxide dissolved in water, to see if the carbon dioxide increased or decreased over time by observing the color change in the solution. A BTB-water solution is yellow when no carbon dioxide is present and blue when the water solution is saturated. If Elodea is photosynthesizing to produce sugar, it uses the carbon dioxide in the water. Therefore, the solution should change from blue to lighter shades of greenish blue to yellow.

Mr. Hopper had his students use the data they collected to write a scientific explanation that responds to the prompt, *Construct a scientific explanation to justify what plants produce and what they use up during photosynthesis*. To respond to this prompt, students needed to use the disciplinary core idea that in photosynthesis, plants use energy from the sun, oxygen, and carbon dioxide to produce sugar and oxygen. But disciplinary core ideas, while essential, are insufficient in responding to the prompt. Students also needed to use the CCC that within a natural system, the transfer of energy drives the motion and cycling of matter. In this case, energy from the sun is transferred and used by plants during photosynthesis. Finally, students needed to use the SEP of constructing a scientific explanation based on valid and reliable evidence obtained from students' own experiments and the assumption that theories and laws that describe the natural world operate today as they did in the past and will continue to do so in the future. Mr. Hopper found that for the most part, his students appropriately constructed the scientific explanations. Still, he wanted to know if students could use their knowledge to make sense of a different scenario in which they had to make use of the same dimensions of scientific knowledge but in a new and different situation.

Mr. Harper had used the *NGSA* design process to create a task that could be used for assessing students as they build toward the unit's focal PE, MS-LS1-6. The task was created using a design blueprint for the learning performance, *Students develop a model that shows that plants (or other photosynthetic organisms) take in water and carbon dioxide to form food (sugar) and oxygen.* Although the SEP was different than what students had used previously in the unit, he thought it would be good to see what students could do using a different SEP but the same DCI and CCC. The task also matched his learning objective of seeing if his students could use the disciplinary knowledge (i.e., the DCI and the CCC) in a new situation using a different practice to make sense of a different but compelling phenomenon. The task he created, *Aquarium Plants and Fish* (see Figure 8.3), would require students to build a model to explain why fish can survive in a closed environment if provided sufficient plant life. In their models, students would need to show how plants use water to generate the necessary oxygen for the fish to survive.

FIGURE 8.3. Mr. Hopper's life science assessment task: Aquarium Plants and Fish

Trevor's teacher set up an aquarium in her classroom, which had a light source, water, plants, and fish. Trevor knows that fish take in oxygen and release carbon dioxide.

Over a month, Trevor observed that the fish are alive and that the plants in the aquarium are growing and have more leaves.

Trevor has learned that plants need food (sugar) to live and develop. Trevor wonders if there is a relationship between how the plants can grow and how the fish get the oxygen they need to live.

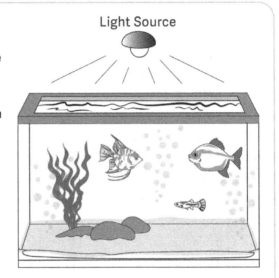

Light Source

1. Draw a model for Trevor that shows the relationship between: (1) how plants can grow and (2) how fish may get the oxygen they need with an aquarium full of plants. Make sure to label all parts of your model.
2. Describe what you have drawn in your model about the relationship between how plants can grow and how fish get oxygen.

The task aligned with Mr. Hopper's classroom instruction, and his class had drawn several models throughout the year, so he felt no need for additional instruction. The next day in class, he challenged his students to draw a model to see if they could apply the knowledge they had been using to explain a different but related phenomenon.

That evening, Mr. Hopper looked through the completed student models. He decided that most of the students could use the idea that plants produce sugar and oxygen from carbon dioxide and water. Yet, some models indicated that students were confused about how additional leaves could grow on the plant over time. He decided this would make a good opportunity for a classroom discussion on the role of photosynthesis and the growth of a plant. The next day, he handed back their models to the students and asked them to use their models to discuss with their partners how plants could obtain materials to grow new leaves. He liked this question because it allowed students to use their prior knowledge of chemical reactions that they had built last year as seventh graders.

Reflection on Scenario 2: Check If Students Can Use Their Knowledge in New Situations

Mr. Hopper used the task *Aquarium Plants and Fish* to see if his class could use the three dimensions of scientific knowledge they developed through his classroom instruction in a new but related situation. The task was a good fit for Mr. Hopper because the task's learning performance provided information about the knowledge and practices students had developed in class. Although the SEP in the performance expectation focused on constructing a scientific explanation, having students draw a model to illustrate the process of photosynthesis seemed appropriate. Mr. Hopper rationalized that using the DCI and CCC with a new SEP would make their proficiency in the three dimensions more useable. As we saw in the first scenario, students need opportunities to learn ideas and different opportunities to build their knowledge. In this scenario, students need to use their knowledge that plants use energy from the sun, carbon dioxide, and water to produce sugar and oxygen. They also needed to apply their knowledge of building models and the flow of energy and matter in an ecosystem to draw a model. The students' models need to show that fish will release carbon dioxide during respiration and that plants will use sunlight and the carbon dioxide the fish release and water to produce oxygen and sugar. Some of the oxygen the plants use for their own respiration, but oxygen is also released into the water.

In this instructional scenario, students collected firsthand data and then used that data to construct a scientific explanation about what plants produce and use up during photosynthesis. Mr. Hopper then wanted to see if his students could apply that knowledge in a new situation with a different phenomenon and scientific practice. The assessment task aligned with his learning goals, allowing students to apply their knowledge.

Notice the difference between the first and second scenarios. In the first scenario, the teacher used a very similar phenomenon to what students were experiencing in instruction. The scenario was very close to the classroom situation that occurred. In the second scenario, the situation was different than what they studied in class—there was no mention of fish using oxygen from water and where the oxygen could come from in a closed system. In this instance, while the scientific ideas are similar, the scenario and SEP differ. The assessment was proximal but not the same as the classroom instruction. Providing these types of tasks puts students in situations where they need to make flexible use of their knowledge, which helps them develop useable knowledge. If we want to support learners in sensemaking and develop useable knowledge of the three dimensions of scientific knowledge, then we need a shift away from using tasks that focus on just recalling what they know and toward students applying their knowledge in similar and proximal situations.

Scenario 3: Encourage Self-Reflection and Dialogue to Promote Learning

The previous scenario explored how Mr. Hopper helped his students develop useable knowledge by having them complete a task in which they needed to use their knowledge of the DCIs, CCCs, and SEPs in a new situation. How else can we support students in developing useable knowledge? Promoting productive classroom discussion and encouraging rethinking to revise one's response can also support the development of useable knowledge. This third scenario illustrates an example from Ms. Daniels' middle school science class making use of a technology-delivered task from the *Next Generation Science Assessment* portal.[1] Let's look at Ms. Daniels' classroom to see how she supported her students to develop useable knowledge through classroom discussion and self-reflection to revise one's position.

Ms. Daniels' unit focuses on how cellular processes transfer energy to living organisms. Throughout the unit, students explored the driving question, *How do I get energy from eating?* Students began the unit by exploring the pathway of food as it moves from the digestive system to the circulatory system and into the cells of the body. Tracing the chemical processes that occur along this pathway, as well as the transfers of energy required. The unit focuses on supporting students in learning the DCI topic of the Organization of Matter and Flow in Organisms (LS1.C), as well as the CCC of Energy and Matter: Flows, Cycles, and Conservation. Ms. Daniels, throughout the unit, also emphasized helping students develop integrated knowledge of the SEPs of constructing and using models and developing evidence-based explanations.

About halfway through the unit, students examined how blood glucose levels changed when animals ingested food, the role of membranes, permeability, and the various inputs and outputs of chemical processes that make energy available to cells. At the end of the unit, students investigated how changes in activity impact the rates of chemical reactions in the body. During instruction, students constructed scientific explanations about where and how food gets used in the body.

As the class approached the end of the unit, Ms. Daniels wanted to see if students could put together the pieces introduced throughout the unit using a single assessment task. Ms. Daniels looked through an online bank of three-dimensional tasks on the *Next Generation*

---

1 The *Next Generation Science Assessment* portal (http://nextgenscienceassessment.org) contains examples of assessment tasks at the elementary and middle school level designed to align with the *NGSS*. The assessment tasks were designed for teachers to use during instruction to gather evidence that students are building proficiency with the *NGSS* performance expectations.

Science Assessment portal (see http://nextgenscienceassessment.org). She selected the task *A Dancer's Energy Use*, shown in Figure 8.4, to use with her students. She thought that this task was a good fit because it used the focal DCIs and CCCs introduced in the unit and it also required that students use a model—in this case, a simulation. The task has the following learning performance associated with it: *Students use a model to explain that in some reactions in an organism, food and oxygen molecules are rearranged to produce carbon dioxide and water and, in this process, energy is released.*

This learning performance helps students build knowledge toward meeting the following *NGSS* performance expectation:

- MS-LS1-7. Develop a model to describe how food is rearranged through chemical reactions forming new molecules that support growth and/or release energy as this matter moves through an organism.

The PE focuses on two disciplinary core ideas:

- LS1.C: Organization for Matter and Energy Flow in Organisms
  Within individual organisms, food moves through a series of chemical reactions in which it is broken down and rearranged to form new molecules, support growth, or to release energy.
- PS3.D: Energy in Chemical Processes and Everyday Life
  Cellular respiration in plants and animals involve chemical reactions with oxygen that release stored energy. In these processes, complex molecules containing carbon react with oxygen to produce carbon dioxide and other materials. (secondary)

The SEP in the performance expectation is to develop a model to describe unobservable mechanisms. Notice how the learning performance, however, focuses on using a model. Ms. Daniels believed that her students needed experience with all aspects of a given SEP to become successful learners. The CCC is Energy and Matter: Flows, Cycles, and Conservation, focusing on how matter is conserved because atoms are conserved in physical and chemical processes. It is the same CCC as found in the learning performance associated with the *Dancer* task.

The *Dancer* task measures students' three-dimensional proficiency and progress by prompting them to manipulate a simulation (SEP: Developing and Using Models) of a dancer by first feeding the dancer a snack and then choosing an activity for the dancer to engage in—e.g., dancing or resting. Next, students "run" the simulation and observe the impact of the selected activity on the rate of production of water, carbon dioxide, and

energy inside the cells of the dancer's body.[2] A line graph depicts the output for each trial or "run" of the simulation. Finally, the prompt asks students to use what they know about the breakdown and rearrangement of matter (DCI) and how energy flows during that process (CCC) to compare what happens when the dancer is at rest versus when they are dancing. Thus, there is a close alignment between the DCI, SEP, and CCC in the task and those addressed in the unit. For Ms. Daniels, the student responses to the task could serve as evidence as to whether they could bring together their varied knowledge from the range of lessons in the curriculum and signal that they were ready to move on to the next unit.

Ms. Daniels found this task valuable to use with her students because it integrated ideas associated with cellular respiration, energy flows, and chemistry in biological systems. Moreover, she knew that her students were interested in movement and dance and so she thought it would be relatable to students' everyday real-world experiences and interests. In completing the task, her students had to make use of the simulation to figure out what happens in one's body when an individual exercises, in this case, dancing. She felt that the task would be motivating and engaging to students and further help them build proficiency in scientific ideas and practices explored in the unit.

Ms. Daniels began the lesson by reviewing the inputs and outputs of cellular respiration using readings and worksheets from the unit materials. In addition, she involved the class in a whole group discussion by asking questions that would highlight how these processes provide energy for living organisms. Ms. Daniels used a variety of discourse moves she learned in a professional learning session (see Miller and Brown, 2019) to steer the group discussion and probe students' integrated knowledge of the three dimensions and to develop their integrated knowledge further. Ms. Daniels focused on students building on each other's ideas. She pressed for evidence from the various firsthand experiences her students conducted in the classroom and from various readings and videos they observed. She asked questions like: "How is your idea like Jamail's?" "Who can add to Guillermo's idea?" "What evidence do you have to support that idea?" "What other ideas can we use?" and "What additional evidence do we need?" Ms. Daniels did not interject any ideas at this point as she focused on the students synthesizing their classroom experiences and bringing forth their knowledge. Ms. Daniels kept track of students' ideas by writing them down on a whiteboard. The professional learning session she attended stressed that for students to build useable knowledge, students involved in classroom discussions needed to build on each other ideas, question the soundness of ideas, and provide evidence to support their ideas.

---

2   You can review the task and run the simulation by going to https://ngss-assessment.portal.concord.org/resources/479/a-dancer-s-energy-use-id-142-03-r02

FIGURE 8.4. Ms. Daniels' online life science assessment task: A Dancer's Energy Use

Jessa is interested in knowing what happens in her body when she is resting and dancing after eating a snack. Use the model to help her figure this out:
1. In the model, press 'Eat Snack' and select 'resting' for the activity level.
2. Observe the graph until the time on the graph reaches 40 seconds.
3. Now, select a new activity level (dancing) and press 'start' to rerun the model.
4. Watch the outputs in the graph until Jessa stops dancing.

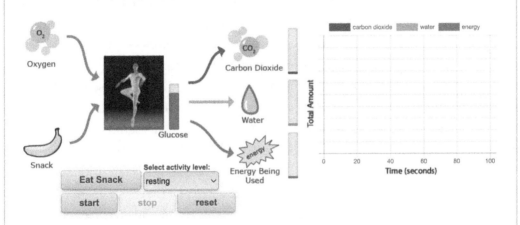

Using the model, describe what will happen to Jessa's levels of water, carbon dioxide, and energy when she is dancing compared to resting. Describe why the levels change.
In your description be sure to include:
- What you know about the breakdown and rearrangement of matter within Jessa's body.
- What you know about how energy is transferred.

After the classroom discussion, she gave students time to work independently on the *Dancer* task. Ms. Daniels provided opportunities for students to think through their responses independently because she felt it helped learners develop their integrated knowledge of the three dimensions and build independent thinking. Ms. Daniels reviewed the responses as students submitted them in the task portal. From their responses, she felt that, although students generally described the changes in the output of the simulation (e.g., "the level of sugar changed because of the dancer's level of activity" and "eating the snack gave her more energy"), they did not describe how the breakdown of glucose and the transfer of energy mediated these changes. She did not know whether this was because students could not explain the change more specifically or because their lack of familiarity with the SEP of Developing and Using Models masked what they knew. To test these two alternative interpretations, Ms. Daniels engaged students in a whole-group discussion where she asked them to elaborate on their responses.

Although the task prompt explicitly asked students to draw on their knowledge of energy transfer in their response (see Figure 8.4), there was little attention to the cross-cutting concept of energy and matter flow in students' responses. For this reason, during the classroom discussion, she purposefully asked students to reflect on how energy and matter flowed through the system. In addition, she focused on having students use the simulation to provide evidence for their responses.

She asked students to share their responses with the class during the discussion. After each response, she asked students to reflect on the response and write in their notebooks if they agreed or not and why or why not. She then called on various students to share their thinking with others in the class and used students' responses in a whole group discussion to probe students for more in-depth responses. She pushed them to use data from the simulation to support their ideas. Once again, she did not interject her ideas because she focused on students reflecting on the responses and asked them their position. She knew if she just told them the appropriate response, students could memorize it, but they would not develop usable knowledge of the idea. She did not evaluate students but asked other students for their reflections on other students' reasoning and the evidence they provided. In this way, she pressed them for further information by using evidence from the simulation and prior classroom work. From the discussion, students provided the specific details about molecular rearrangement and energy transfer that she hoped students would develop. Notice how Ms. Daniels did not use the tasks and students' responses for evaluative purposes but to push forward their learning. This discussion showed that although students could respond superficially, they did not demonstrate the full extent of their knowledge while individually completing the assessment task.

Reflection on Scenario 3: Encourage Self-Reflection and Dialogue to Promote Learning

Given how students responded to the prompts on the tasks and what she hoped her students would learn, Ms. Daniels decided to use the task as a tool for self-reflection and to bring out students' ideas through classroom discourse. She decided that the class should spend more time discussing their ideas and reflecting on their responses to the prompts. She knew that to support students in developing integrated knowledge of the three dimensions, they needed to reflect on their responses, provide evidence for their responses, and question the ideas of others to see how they fit or do not fit with their ideas. These reflections would support them in synthesizing the complex ideas associated with the learning performance and performance expectation.

Note that Ms. Daniels' goal focused on students going beyond using the disciplinary core ideas that the LP focused on by using evidence obtained from modeling. To do so, she asked students to use evidence from using the simulation, a model, to support their response. In this way, she also obtained evidence of students developing knowledge of the SEP associated with the learning performance. She also saw a chance for deeper engagement with the CCC of Energy and Matter: Flows, Cycles, and Conservation. She asked questions that would allow students to bring forth their ideas, once again having students use evidence from the simulation to support their claims.

This scenario demonstrates how a teacher can use a three-dimensional task, like the *Dancer* task, along with self-reflection and classroom discussion using appropriate discourse moves to create expanded opportunities for students to demonstrate the full extent of their integrated knowledge of the disciplinary core ideas, crosscutting concepts, and science and engineering practices. Notice that Ms. Daniels needed to know how to conduct a classroom discussion using various discourse moves. These discourse moves allow students to synthesize and reflect on their thinking. However, many teachers may find these discourse moves challenging to use as they require a teacher not to give the correct response or to evaluate student responses but rather to bring out and guide thinking and reasoning by considering other students' perspectives and the evidence that students bring forth. The teachers' primary responsibility is to guide the discussion in a manner that will allow students to bring forth their ideas and the evidence they have for supporting them. As we see from this scenario, Ms. Daniels also supported students synthesizing their ideas by recording them on a whiteboard.

Primary Takeaways

In this chapter, classroom scenarios illustrate how teachers can use assessment tasks in different ways to learn about students' performance to support their three-dimensional learning and what teachers can do to help students further develop their integrated knowledge of the three dimensions of scientific knowledge. We discussed the opportunities for each way of using tasks. The three primary strategies of how to use assessment tasks focus on supporting student learning in instructionally supportive ways, reflecting current evidence-based teaching practices for classroom assessment (Stiggins, 2014). In each scenario, teachers used a task for a different purpose, but in each case, the use of the task was to promote student development of the three dimensions of scientific knowledge.

Ms. McGinnis used the task *Which Metals are the Same?* to check if students built their three-dimensional knowledge sufficiently. Classroom instruction and assessment

tasks need to support students in building proficiency toward performance expectations. Therefore, classroom instruction and assessment tasks work together synergistically to help learners construct knowledge of the three dimensions of scientific knowledge and provide teacher and student feedback about their learning. Mr. Hopper used the task *Aquarium Plants and Fish* to check if students could use their knowledge in a new situation. He checked if students could use their knowledge with a new phenomenon and with a different SEP (i.e., whether they developed knowledge-in-use) by involving them in completing a task in which they needed to use their knowledge of the disciplinary core ideas, crosscutting concepts, with a different science and engineering practice, to make sense of new phenomenon to which the students had not previously applied their knowledge. Ms. Daniels used the *Dancer* task to see if students had developed the three-dimensional knowledge the unit focused on by encouraging self-reflection and dialogue. Promoting productive classroom discussion and encouraging rethinking to revise one's response can support the development of useable knowledge. We provide key takeaways for using assessment tasks in instructionally supportive ways in Table 8.1.

TABLE 8.1. Strategies With Key Takeaways for Using Assessment Tasks Within Instruction

Strategies to Use Assessment Tasks to Support Learning	Key Takeaway
1. Follow students' progress as they build proficiency toward three-dimensional learning goals	Classroom instruction and assessment tasks need to support students in building proficiency toward performance expectations. Classroom instruction and the assessment tasks work together synergistically to help learners construct knowledge of the three dimensions of scientific knowledge and to provide feedback to the teacher and student about their learning.
2. Check if students can use their knowledge in new situations	Teachers can check if students can use their knowledge in new situations (i.e., whether they developed knowledge-in-use) by involving students in completing a task in which they need to use their knowledge of the disciplinary core ideas, crosscutting concepts, and science and engineering practices, but in a new situation and/or with a new phenomenon to which students had not previously applied their knowledge.
3. Encourage self-reflection and dialogue to promote learning	Promoting productive classroom discussion and encouraging rethinking to revise one's response can also support the development of useable knowledge.

These three scenarios also show the teachers embedding the tasks at different junctures in their curricular sequence, with differing goals for using the assessment tasks. While the three cases show different ways to use assessment tasks to promote three-dimensional learning and instruction, many other possible ways exist to accomplish the same goal.

We invite you to talk through these three strategies of using assessment tasks to support student learning with your colleagues to determine when and how to use classroom-based assessment tasks to support the development of three-dimensional knowledge in your classroom. We typically think of using assessment tasks for evaluation, and while evaluation is an important task a teacher needs to perform for accountability reasons, the power of embedded classroom assessment is to provide the teacher and students with information regarding student thinking and performance to promote three-dimensional learning.

Finally, you might find the following questions as productive starting points for self-reflection and discussion with colleagues about how to plan and enact instruction and assessment in support of three-dimensional learning.

Questions to Guide When You Will Use the Task Within an Instructional Sequence:

a. What DCIs, SEPs, and CCCs do this lesson, series of lessons, or unit support? What are the dimensions that are the focus of this task and how do they relate to your instruction?

b. What does this task measure, and how does it relate to your evidence statement—the observable evidence you are trying to gather about what students know and can do?

c. When in your instructional sequence will it be most advantageous to use this task to support student learning?

d. What will student responses on this task tell you about their integrated knowledge of the three dimensions with respect to instructional goals?

Questions to Help You Consider How You Will Use the Task With Your Students:

a. What are the critical connections between the three dimensions of the assessment task and the three-dimensional goals of your instruction?

b. How do you foreground the critical connections both before students complete this task and while they are completing it?

c. How can you scaffold students' experiences with the dimensions (the DCIs, SEPs, and CCCs) within instruction, so they will bring what they already know to complete this task?

Questions to Guide What You Will Do Next With Student Responses:

a. What DCIs, SEPs, and CCCs might students need additional support for or practice with? Where can you build opportunities to do that?

b. What discourse moves can you use to nudge students to reflect on whether they presented sufficient evidence and appropriate scientific reasons to support their claims?

c. What new instructional experiences do students need to advance their thinking about how to use the three dimensions to make sense of a phenomenon or solve a problem?

d. How might I modify the scenario to better relate to my students' interests to further motivate them to use the three dimensions to make sense of phenomena?

References

Krajcik, J. S. & J. Nordine. 2016. Energy in photosynthesis and cellular respiration. In *Teaching energy across the sciences, K–12*, ed. J. Nordine, 79–103. Arlington, VA: National Science Teachers Association Press.

Miller, E. & T. Brown. 2019. Discourse tools amplify students' ideas and build knowledge. *Science and Children*, 56(9), 76–79.

National Research Council (NRC). 2012. *A framework for K–12 science education: Practices, crosscutting concepts, and core ideas.* Washington, DC: National Academies Press.

NGSS Lead States, 2013. *Next Generation Science Standards: For states, by states.* Washington, DC: National Academies Press.

Stiggins, R. 2014. *Revolutionize assessment: Empower students, inspire learning.* Thousand Oaks, CA: Corwin Press.

CHAPTER 9

Reflections and Implications: Creating and Using Three-Dimensional Assessment Tasks to Support *NGSS* Instruction

Christopher J. Harris, WestEd • Joseph Krajcik, Michigan State University • James W. Pellegrino, University of Illinois Chicago • Daniel Damelin, Concord Consortium

Our central aim in writing this book has been to provide a systematic process for creating assessment tasks consistent with today's vision for science education. This vision emphasizes that to truly know and understand in science is to be able to use the three dimensions of scientific knowledge to explain compelling phenomena and provide solutions to complex problems (National Research Council, 2012; NGSS Lead States, 2013). In science classrooms, this means that it is not solely what students know, but also how they use and apply what they know that really matters for science learning. Students learn by applying the three dimensions of science and engineering practices, disciplinary core ideas, and crosscutting concepts in the context of reasoning, sensemaking, and problem solving. Through three-dimensional learning, students actively use their knowledge and over time become more adept at applying their knowledge to new situations or problems that relate to what they have previously learned. Importantly, using and applying knowledge transforms that very knowledge: It becomes more robust and deepens students' proficiency with science as both a body of knowledge and a way of understanding the world. The notions of knowledge-in-use and three-dimensional learning are main features of the contemporary vision for science education and considered central to teaching and learning, and in assessing what students know and can do.

This remarkable vision for science education has changed classroom instruction and assessment. For instruction, the *Framework* and *NGSS* have sparked a new era of curriculum materials aligned to the vision and to three-dimensional learning goals. Along with curriculum materials, new professional learning resources and programs have been designed and made widely available to teachers. In science classrooms in many corners of the country, teachers and students are shifting instruction to fit a three-dimensional view of science learning. In these classrooms, for example, students go beyond just learning about a science topic and toward figuring out and explaining how or why something happens in the natural world. Teachers use phenomena to drive instruction and create meaningful situations for three-dimensional learning to occur. Importantly, teachers activate, monitor, and support students' three-dimensional learning.

For assessment, the *Framework* and *NGSS* have placed the spotlight on classroom-based assessments that can be used formatively to guide instruction and provide feedback to individual students or the whole class as they build their science proficiencies. This dramatically changes the role of classroom assessment away from solely an evaluative role where assessment is used primarily for grading purposes and toward an instructionally supportive role where assessment is used for improving teaching and advancing three-dimensional learning. To make this change a reality, assessments need to be designed and used to measure what matters for science instruction. In recent years, good progress has been made in figuring out how to create three-dimensional assessment tasks. Yet today, there continues to be a pressing need for classroom-based assessment tasks that help students put their knowledge to use, instead of just requiring the recall of that knowledge. At the same time, there continues to be a significant need for science educators to learn how to create tasks that they can use to gain insight into students' progress in building knowledge toward all that is required by the *NGSS* performance expectations. This book offers the *Next Generation Science Assessment* design process as an innovative solution to addressing these important needs.

In this closing chapter, we revisit the *Next Generation Science Assessment* (*NGSA*) design process with an eye toward bringing forward the benefits of using the process for deepening one's professional knowledge for teaching and fulfilling the important requirements for assessing three-dimensional learning. We also offer practical guidance for how teachers can take their first steps with using the *NGSA* design process. We then consider future directions for classroom-based task design and in particular, the emerging role of technology for empowering teachers and students to use assessments in innovative ways. The chapter concludes with thoughts about the value of bringing science instruction and assessment closer together through the *NGSA* design process.

Revisiting the Value of the *Next Generation Science Assessment* Design Process

The *NGSA* design process equips science educators—including teachers, instructional leaders and coaches, professional learning facilitators, state and district educational leaders, and curriculum and assessment developers—with a systematic approach to create valuable assessments that support instructional practice and students' three-dimensional learning. The process uses *NGSS* performance expectations as the starting point for creating sets of learning performances that guide the development of assessment tasks and rubrics. Learning performances are intermediary three-dimensional performance targets for instruction and assessment that can signal whether students are moving along a productive path to proficiency with performance expectations (PEs). In the *NGSA* process, learning performances are a keystone for developing tasks that can be used during instruction to assess "building toward" the knowledge-in-use learning goals of the *NGSS*.

As introduced in Chapter 2, the *NGSA* design process involves six major steps across three phases. The first phase, Steps 1–3 described in Chapter 3, begins with selecting a PE or PE bundle and systematically unpacking the dimensions to understand the assessable components. The elaborations from the unpacking are used to create a visual representation in the form of a map, which we refer to as an *integrated dimension map*, that lays out the dimensional terrain for fully achieving the PE or bundle. An integrated dimension map describes key relationships among the elements of the disciplinary core ideas and identifies how aspects of the science and engineering practices and crosscutting concepts can work with these disciplinary relationships to promote students' integrated proficiency. The second phase, Step 4 described in Chapter 4, entails using the integrated dimension map to articulate and refine a set of learning performances that collectively describe proficiencies that students need to demonstrate to meet a PE or PE bundle. Each learning performance is written as a three-dimensional statement that represents some portion of the terrain of the PE or bundle. The third phase, Steps 5–6 described in Chapters 5, 6, and 7, calls for using an organizational strategy called a *design blueprint* to guide the principled development of tasks and rubrics for each learning performance. Every design blueprint includes five key elements for designers to do their work: integrated proficiencies, evidence statements, essential task features, variable task features, and equity and inclusion considerations. The final step in the design process involves using the blueprints to create tasks and accompanying rubrics. Essentially, the blueprints serve as the ground plans for task designers, providing clear information for creating tasks and reference points for checking work.

Highlighted in Chapter 6 are design principles and considerations for ensuring that tasks are accessible and fair for a wide range of students with varying backgrounds, skills, and abilities so that they can demonstrate three-dimensional learning.

If you follow the *NGSA* design process, you will create tasks that are well-aligned with three-dimensional learning goals. Moreover, you stand to create tasks with phenomena and scenarios that relate to your students interests and curiosities about the world. These tasks will allow multiple ways for students to demonstrate their developing proficiencies and support engagement so that students will be more likely to persist in reading and responding to them. *Is it worth the extra time and effort to create tasks that will make such a difference for students?* We believe yes, and hope that you agree. As we wrote in Chapter 1, much of typical assessment development relies on intuition and "miracles" in going from standards to assessment tasks that can be used in the classroom to inform teaching and learning. If you follow the process described in this book, you can make substantial progress in creating valid assessments of students' knowledge and capabilities associated with the *NGSS* performance expectations and these assessments can be instructionally useful and usable.

What You Learn by Following the *NGSA* Design Process

Important to consider, too, is what you can gain in your professional knowledge by following the *NGSA* design process. Each phase of the process provides an opportunity for learning that will increase your knowledge for three-dimensional instruction and assessment. Below, we briefly describe the benefits that science educators gain when immersing themselves in five major activities of the design process: unpacking and mapping the *NGSS* dimensions, constructing learning performances, attending to equity and inclusion, specifying blueprints and task features, and constructing tasks and rubrics.

Unpacking and Mapping the NGSS *Dimensions*

The *NGSA* design process can be used to unpack and map the meaningful parts of the PEs that will be suitable for classroom-based assessment. Unpacking and mapping is of high value because it enables you to identify and elaborate on all that is involved for students to successfully demonstrate integrated proficiency with PEs. It focuses your attention on the specific and meaningful aspects of the three dimensions as well as the knowledge and capabilities that students need to develop for each of them.

Unpacking pushes you to have your students in mind right from the start of the design process by requiring that you describe students' prior knowledge and identify likely student challenges with the dimensions; define boundaries of what students should know and be able to do; earmark issues of equity and inclusion that are relevant to the dimensions; identify candidate phenomena relevant to both the PE and students' everyday lives and interests; and sketch out possible realistic scenarios that can provide a motivating context for making sense of phenomena. Mapping then synthesizes selected information from unpacking in a visual representation, called an *integrated dimension map*, that shows the area and boundaries of a performance expectation or bundle. Mapping is beneficial for creating a clear picture of what needs to be addressed for achieving the PEs. Together, unpacking and mapping allows you to develop rich knowledge of the PEs in a manner that deeply connects to your classroom instruction and assessment needs.

Constructing Learning Performances

The *NGSA* design process emphasizes using the meaningful parts of PEs to construct comprehensive sets of smaller performance statements that are called *learning performances*. Learning performances are knowledge-in-use statements that take on the three-dimensional structure of a PE but are smaller in scope and align to only a portion of the PE. Each learning performance describes an essential part of a PE or PE bundle that students need to demonstrate at some point during instruction to show that they are making reasonable progress toward mastering all the knowledge elements that underlie a single PE or PE bundle. Collectively, they are keystones for developing tasks that enable you to assess the progress of your students in building knowledge toward mastery of the PEs.

Constructing learning performances provides a valuable opportunity to learn firsthand how to write three-dimensional performance goals. It takes time to become proficient in constructing them, but in developing your proficiency you will also deepen your integrated knowledge of the dimensions and how they can work together. Importantly, if you go through this process, you will come to appreciate what it takes for students to use their knowledge to make sense of phenomena or solve problems. With learning performances in hand, you can create tasks that will bring out three-dimensional performance in your students. Finally, constructing learning performances can help you identify important opportunities that you should attend to and assess *before* the end of an instructional sequence or unit. This is highly valuable for instructional planning.

Attending to Equity and Inclusion in the Design Process

Designing and enacting three-dimensional assessment tasks that promote equity and inclusion are a responsibility of all science educators. From start to finish, the *NGSA* design process takes you down a path toward developing tasks that will enable all your students to show how they can use and apply what they know. The process deliberately encourages you to consider how assessment design can be made relevant to your science instruction and your students.

The guidance that we offer for task design includes using scenarios where the phenomena will be of interest and have relevance for students, explicitly attending to language so that students will be more likely to persist in reading and responding to tasks, embedding scaffolds to make expectations explicit for students, and varying the ways across tasks for students to demonstrate their performance. As you follow the process over multiple task development cycles, you will become more attuned to the varying backgrounds, skills, and abilities of your students and adept at leveraging students' background knowledge and experiences in task design. For example, it can initially be challenging to create the overarching context or situation that frames each task so that it is relevant to science and/or engineering design while also relatable to students' everyday real-world experiences or interests. By engaging with the design process, you will come to know assessment techniques and formats that will work best in allowing your students to demonstrate a range of evidence of their three-dimensional learning.

Specifying Blueprints and Task Features

Design blueprints are an organizational strategy for the principled development of tasks aligned to learning performances. They serve as the ground plans for task design and set the boundaries for what designers should include or not include in tasks. They also ensure that critical specifications like task features, scaffolding levels, and format types are used consistently. Simply put, blueprints are important organizers that encourage you to consider what is essential in tasks and establishes consistency in design.

When specifying blueprints, you will need to draw from all your prior work in the design process to identify the "must haves" for a well-designed and aligned task. Creating them is valuable because they help you bring together all the technical information so that you can begin to see the possibilities for constructing tasks. Once developed, design blueprints spark the creative process: they can be used by you and even shared with colleagues to develop one or more tasks that will align with a learning performance. A great benefit of a single blueprint is that it can be used to create multiple tasks that all share essential attributes while also varying on some features. This allows you to create a set of tasks that vary in some useful ways yet still maintain alignment with the learning performance so that you assess what matters for instruction.

Constructing Tasks and Rubrics

Design blueprints provide essential technical information, but not to be overlooked is the role of creativity in bringing everything together into a well-constructed task with an accompanying rubric that will assess meaningful performance. Assessment tasks for three-dimensional learning should be designed to be intriguing and relevant to students and robust enough so that students can showcase their three-dimensional performance. A well-designed task will have as the centerpiece a phenomenon that is presented in a scenario that creates a real-world need for using and applying knowledge to make sense of the phenomenon. Notably, a good phenomenon and surrounding scenario will set the stage for students to respond to the task by creating a context that leads students to encounter and grapple with the essential aspects of the learning performance. A well-designed rubric will illustrate what three-dimensional performance looks like. It will support you to make accurate and equitable judgments about student performance, differentiate varying levels of performance among students, and formulate actionable student feedback.

By following the *NGSA* design process and using it over time, you will deepen your expertise in constructing three-dimensional tasks and rubrics. An added benefit is that you can then apply your expertise to evaluate already existing assessment resources to determine whether they align with *NGSS* three-dimensional learning and then modify them as needed. Using tasks and rubrics in your classroom will also increase your knowledge about how to elicit and interpret students' three-dimensional performance. This knowledge can be invaluable for instructional planning. In these ways, constructing and using tasks and rubrics can help you innovate your instructional and assessment practices and enrich your understanding of what is needed to actively support three-dimensional learning.

Finally, the *NGSA* design process encourages you to design with your students in mind. The tasks you develop for your students will have meaningful phenomena and problems presented in scenarios that provide important practice for knowledge-in-use. As you become more expert in three-dimensional task design, you will learn how to provide equitable opportunities within tasks to help students make sense of what they are asked to do. Moreover, the varied features in these tasks will enable you to reach a wider range of students because the tasks will encourage use of multiple modes to demonstrate performance, such

as through science and engineering practices that support expressing three-dimensional learning in different ways. Allowing for multiple ways for students to demonstrate what they know and can do opens greater opportunities for you to gain insight into students' performance and use the information to improve teaching and advance learning. Rubrics can help you make sense of students' responses and chart evidence of how students are progressing in building three-dimensional proficiency.

Getting Started

The *NGSA* design process is meant to prepare you to develop and use three-dimensional assessments that will help your students become engaged and active in using and applying what they know. As you get started, the excitement for following the design process can be especially strong. Yet, it is important to recognize that this is a very different process from the typical design approach used for classroom assessment that we described in Chapter 1. Many who follow the *NGSA* design process for the first time find that it takes time, effort, creative thinking, and practice—all that one might expect to be required when learning a new complex skill. With this in mind, we offer some practical guidance for launching the design process in a measured way so that you can build your expertise over time.

Start on a Small Scale

Begin with one performance expectation or a manageable bundle of two closely related performance expectations and treat this first undertaking as a learning opportunity. Starting on a smaller scale allows you to familiarize yourself with the process, reflect, and make adjustments that you can use in your next design cycle. We recommend that you select a performance expectation or bundle that falls within a disciplinary topic that you have familiarity with, represents a robust learning goal for your students, and that aligns strongly with your instruction. Alignment to instruction is essential for ensuring that the tasks you design will produce relevant information at appropriate times to support instructional decision-making. Once you get the process underway, we suggest you proceed in a deliberate and manageable way keeping in mind that you are learning yourself as you work though each phase. Be sure to recognize that it takes time to learn the process and that you will get better over time in your design work. Recognize, too, that engaging in the process in a thoughtful way will deepen your understanding of all three dimensions. For instance, *if you do not know crosscutting concepts very well, you will after going through this process!*

Use the Process Strategically

This is not a process for creating one-off tasks that will only be used a single time. The *NGSA* design process is best suited for designing tasks that can be used over and over again to support instruction across classes and years. Moreover, the assessment resources you develop including integrated dimension maps, learning performances, and design blueprints can be used time and again in varying ways to support assessment and instruction. For example, integrated dimension maps can be used to gain a global view of all the relationships in a performance expectation or bundle which in turn can be used for both instructional and assessment design. Learning performances can be used in the instructional planning process to plot how students will be expected to build proficiency toward a performance expectation or bundle. They can also be used to identify junctures in an instructional sequence where it will be beneficial to assess students. Moreover, you can use learning performances to plan instructional activities. If you use learning performances to develop instructional activities and assessments, you will have strong alignment with what students learn and what you assess.

Another strategic use to keep in mind is that design blueprints can be used by individual teachers or shared among colleagues, such as within a grade level team, to construct multiple tasks that all align to the same learning performance but vary in features such as scaffolds, scenario contexts, and modalities for students to demonstrate proficiency. These are just a few examples of the many ways that the resources produced from the design process can be used strategically to benefit your teaching and promote student learning. As you immerse yourself in the design process, we encourage you to consider at each phase how you can maximize the usefulness of the resources you are developing.

Apply a Design Thinking Mindset

You likely noticed that the *NGSA* design process proceeds in a linear fashion, with earlier steps informing later steps. This structure helps you progress from performance expectations which are the end goals of instruction, to learning performances that describe the intermediate performances that students need to demonstrate on their way to achieving the end goals, and then to tasks and rubrics that can be used during instruction to assess students as they build toward the performance expectations. At first glance, the process can appear mechanical where you move lockstep through each phase. In practice, it is very much a creative and dynamic activity that will enable you to bring together your expertise with varied knowledge from different sources to develop innovative assessments for classroom use. In your role as a designer and implementer of assessment, it can be very

helpful to take on a mindset where you use the design process to innovate and generate new knowledge and make insights that will advance your assessment design work. This means being flexible with the process—allowing yourself, for example, to go back a step to revise based on what you have learned in a later step—and encouraging yourself to be a learner in the process. With time, practice, and a designer mindset you will become proficient and confident in task and rubric design.

Design, Implement, and Reflect With Colleagues

There can be tremendous value in collaborating with colleagues in the *NGSA* design process and in the sharing of tasks and resources that have been developed through the process. Working with colleagues can help you advance in your design work and deepen your design knowledge as it provides ongoing opportunities for you to jointly design, discuss, and revise as you proceed through each phase of the process. Potential benefits of collaborative planning and ongoing collaborative conversations include changes in both understanding of assessment design and assessment practice over time, and the development of shared strategies for dealing with the practical challenges of designing and implementing three-dimensional tasks. When getting started with the design process, a perk of working together is that it can make your first run through the design process less daunting since you can support and coach each other along the way. A strategy that some teachers find beneficial is to undertake a cycle of collaboration that begins with collaborative design among colleagues to construct a set of tasks that match well with instruction, followed by teachers implementing the assessment tasks with their students and using rubrics to interpret students' responses, and then reflecting on the enactments with one another and deciding on next steps for instruction. A further benefit of undertaking cycles of collaboration is that it can encourage you and your colleagues to move away from using traditional or narrow assessment techniques and formats and toward trying new ways for using assessment to support three-dimensional learning. A final benefit includes sharing the unpacking documents, learning performance, dimension maps, blueprints, tasks, and rubrics with others. You will certainly end up with more useful tasks working with colleagues than by just working by yourself.

Future Directions: The Role of Technology

Increasingly, schools are transitioning to digital teaching and learning tools where each student has an electronic device to access the internet and interact with digital learning materials and teachers have digital access to support their teaching. We now live in an era where wireless and networked technologies are becoming deeply woven into classroom life and there is exciting potential for technology to empower teachers and students to use assessments in innovative ways. In the not-too-distant future, the benefits of technology-delivered assessments that can support three-dimensional learning will be realized on a regular basis in classrooms. Moreover, new design applications will soon become available to enable teachers to efficiently design and implement their own assessment tasks that measure the active and integrated learning that comes from robust science instruction.

Technology-delivered assessments have a multitude of advantages for teachers and students. For students, technology enhancements such as video and simulations can expand the phenomena that can be investigated. Various assistive technologies make assessment materials more accessible to students, for example through screen readers that facilitate navigation and reading of text and speech-to-text capabilities that support students in responding to tasks. By providing background drawings, drawing tools, stamps, and modeling features, technologies help scaffold students in demonstrating their learning in deeper ways. Moreover, because technology-delivered assessment tasks can enable students to use multiple modalities and representations, students with diverse abilities and language backgrounds may have better opportunities to demonstrate their proficiency than possible in print-based assessment. For teachers, technology is well-suited to support implementing assessments with relative ease and for collecting student response data. Moreover, recent advances in automated assessment technologies such as machine learning can help teachers more readily interpret and use assessment information to improve teaching and advance learning. This line of work is already underway and changing the boundaries for what assessment can look like in science classrooms. As these tools become more widely available, we encourage you to critically examine and try out promising technologies that stand to enhance the design and implementation of assessment consistent with today's vision for science education.

On a related note, many of the authors who wrote this book have ongoing work focused on developing technology-enhanced science assessment tasks with rubrics and accompanying digital instructional resources for elementary and middle school classrooms. Some of their assessment tasks and resources have been made available to science educators through the *Next Generation Science Assessment* online portal (see http://nextgenscienceassessment.org). The portal contains examples of assessment tasks at the elementary and middle school level designed to align with the *NGSS*. The assessment tasks were designed for teachers to use during instruction to gather evidence that students are building proficiency with the *NGSS* performance expectations.

Final Thoughts

This book has been written for those concerned with the design of science assessments for classroom use with an emphasis on *using assessment for three-dimensional learning*. The kind of tasks that are needed for today's science classrooms must assess the active and integrated learning that comes from instruction that engages students in using and applying multiple dimensions. To accomplish this, science educators need to design tasks that will support students to actively use their knowledge—instead of just recalling that knowledge—to make sense of compelling phenomena and design solutions to complex problems. The value of useable knowledge is twofold. First, by creating assessments that require students to use their knowledge, we help them to understand when, how, and why to apply what they have learned. Second, we help them develop the integrated and transferrable knowledge and skills needed for future learning, problem-solving, and decision-making.

Within this book we presented comprehensive guidance called the *Next Generation Science Assessment* design process to enable science educators to design assessments that align with the contemporary vision for science education and are useful and usable in the classroom. The process is a systematic, multi-step approach to designing assessment tasks that integrates disciplinary core ideas, science and engineering practices, and crosscutting concepts as called for in the *Framework* and the *NGSS*. The process can be used to create tasks that are very different from traditional formats of assessment. Using this process will enable you to develop the knowledge and capabilities to design a variety of tasks that can provide a picture of how students' three-dimensional science learning builds over time with your guidance. Through this process, you will also learn the real value of bringing instruction and assessment closer together which is to support meaningful and equitable science learning opportunities for all the students you teach, helping to create a more just and sustainable society.

References

National Research Council (NRC). 2012. *A framework for K–12 science education: Practices, crosscutting concepts, and core ideas*. Washington, DC: National Academies Press.

NGSS Lead States. 2013. *Next Generation Science Standards: For states, by states*. Washington, DC: National Academies Press.

Acknowledgments

THIS WORK WOULD NOT HAVE BEEN POSSIBLE without the high-caliber contributions of many science educators including teachers, school leaders, and early career science education researchers and science assessment experts, as well as the many students who participated in the piloting of tasks as we sought to establish their validity and usability in classrooms. Science teachers were instrumental in helping us develop the *NGSA* design process and validate the products of that process for classroom use. Core collaborative team members who have worked with us over the years made contributions to this book and are listed as chapter authors. In addition to them, there were many other project team members who worked with us at various stages including Angela DeBarger, Louis DiBello, Reina Fujii, Mon Lin Ko, Kiley McElroy-Brown, and Gauri Vaishampayan, among others.

Important to acknowledge are science education experts and stakeholders who were not part of the development team but served as advisors to our work and sat on expert review panels. Individually and collectively, they provided invaluable feedback at many points along the way. These thought partners included Bruce Alberts, Carlos Ayala, Aneesha Badrinarayan, Derek Briggs, Angela DeBarger, Dawn O'Connor, Melanie Cooper, Kate McNeill, Stephen Pruitt, Trish Sheldon, Maria Simani, and Ted Willard among others.

In bringing this book together, we gratefully acknowledge the excellent support provided by Elizabeth Arnett and May Chan in preparing the chapters for publication as

ACKNOWLEDGMENTS

well as Henrique Cirne-Lima and Lucia Yuan in preparing graphics. May Chan, in particular, worked closely with us and provided outstanding editorial support that elevated the quality of the book. We thank all the reviewers, including those who were enlisted by NSTA Press and those colleagues who were recruited by us at various stages in the writing and editing process, for their thoughtful feedback on the chapters.

We also extend thanks to our major funders—the National Science Foundation, the Gordon and Betty Moore Foundation, and the Chan Zuckerberg Initiative—whose generous contributions made this work possible.

Challenging problems require the collaboration of individuals with diverse skills, perspectives, and backgrounds. This project brought together such a team to develop a vision of what classroom-based assessment tasks that align with the *NGSS* could look like and how teachers could use them in the classroom. It is the combined vision, expertise, and dedication of this team that allowed for the development of this work. We are grateful to all the individuals who helped shape our vision of classroom assessment and pushed forward our understanding of the *Framework* and *NGSS*.

This material is based upon work supported by the National Science Foundation (Grant numbers 1316874, 1316903, 1316908 & 1903103), the Gordon and Betty Moore Foundation (grant number 4482), and the Chan Zuckerberg Initiative (grant number 194933). Any opinions, findings, and conclusions or recommendations expressed in this book are those of the authors and do not necessarily reflect the views of their own institutions, the National Science Foundation, the Gordon and Betty Moore Foundation, or the Chan Zuckerberg Initiative.

INDEX